...Glass is more gentle, graceful, and noble than any metal and its use is more delightful, polite, and sightly than any other material at this day known to the world.

Antonio Neri, 1612

A SHORT HISTORY OF GLASS

Chloe Zerwick

The Corning Museum of Glass
Corning, New York

Cover: Aquatint after J.M. Volz by C.
Meichelt from *Glashütte im "Aule"*
*(Early Nineteenth Century Glassmaking in
the Black Forest)* by A. Schreiber.
Freiburg, 1820-1827.

Photography: Raymond F. Errett and
Nicholas L. Williams
Art Direction: Anthony Russell
Design: Peg Patterson
Typography: Volk & Huxley
Printing: Froelich/Greene Litho Corp.

CONTENTS

FOREWORD

People have been making glass for at least thirty-five centuries. About 1500 years before the birth of Jesus, back when iron was first used, back when Moses led the Israelites out of Egypt, glass was new. That was a very long time ago—in fact one hundred and seventy-five successive generations have gone by. Since those ancient days, uncounted thousands of people have worked with glass, people in almost every part of the world, people with traditions, needs, and ideas as varied as history itself. To discover and bring all this human activity together is the job of The Corning Museum of Glass: we do it by collecting and analyzing evidence. The best source is the glass object itself because each piece is like a time capsule reflecting the ideas of its maker, of the techniques, practices, and styles alive at the very moment in which it was made. So far we have acquired by gift and purchase more than 19,000 glass objects. Archeology gives us more evidence, recording as it does the lives of people, of communities, of civilizations in which glass has played a part. But the source that multiplies the fastest is the printed word. The Museum's Library now contains more than 25,000 publications; by the year 2000 it expects to house 51,000, more than double everything ever printed on the subject before!

This concise history is a summary, largely visual, of what we now know of all the centuries of experience, of the endless technical discoveries, of the limitless artistic inventions of generation after generation. Wherever possible, words written during the time when the glass was made are quoted. These, like the pieces themselves, are direct evidence; they give a lively idea of just what various people thought about glass throughout history.

As a review of the most important glass in the collection, this book is also a simplified summary of the Museum's first three decades. After thirty years of looking, collecting, exhibiting, publishing, and looking again, we know that glass has magic about it, that it seems to fascinate both makers and owners regardless of where or when they live. Perhaps it is because glass is molten and moving and untouchable in the very moment of its formation that the craftsman must respond with great speed and certainty. Perhaps it is its split nature— opaque or transparent, colored or clear, fragile or strong. Perhaps it is its seemingly endless potential. With glass now in the hands of artists as well as craftsmen, with wires of glass threads transmitting light impulses across the land, with eyeglasses that lighten and darken with the sun, we may be seeing— within all the wonders of this long history—just the beginning.

Thomas S. Buechner
Director
The Corning Museum of Glass

INTRODUCTION

Glass: Shiny, hard, fragile — shattering in an instant or surviving for thousands of years — a rigid liquid that is worked in a molten state — too hot to touch, yet often made by hand — molded, blown, cut, engraved, enameled, or painted. Of the craftsman, it demands the ultimate in steady nerves, skill, control, patience, judgment, and spontaneity. To possess such rigorous qualifications requires not only native gifts but long experience, so it is no wonder that the head glassmaker is called a "gaffer," an old word for grandfather. The gaffer's reheating furnace is called a "glory hole," a tribute to the beauty he creates in the fire.

Though some glass objects have been as precious as gold, glass can be formed by melting together the most ordinary natural materials, found nearly everywhere: sand, the main ingredient of glass; ashes (alkali), usually made by burning certain plants or trees, to make the sand melt at a feasible temperature; and a stabilizing substance such as lime, made from crushed stones, to protect the glass from the attack of moisture. Virtually the same glass recipe as that given on a cuneiform tablet of the 7th century B.C. is in use today.

The invention of glass more than 3500 years ago is shrouded in uncertainty. The Roman historian Pliny attributed it to Phoenician sailors. He recounted how they landed on a beach, propped a cooking pot on some blocks of natron (an alkali) they were carrying as cargo, and made a fire on which to cook a meal. To

From Diderot, et al, *Encyclopedie.* 18th century.

their surprise, the sand beneath the fire melted and ran in a liquid stream which then cooled and hardened into glass. Hardly anyone believes Pliny's story. Others say glass evolved from the manufacture of faience, an older material with a white interior and a colorful, shiny surface. Faience can be made at a lower temperature than glass and is composed of crushed quartz and alkali. These same ingredients, mixed in slightly different proportions, form a true glass if subjected to a higher temperature.

However, no one really knows how glass came to be made. It is older than the Ten Commandments and probably originated somewhere in the Middle East. It was adopted and importantly developed by the Romans, flowered under the Islamic empire, reached new heights in Renaissance Venice from whence it spread throughout Europe and, eventually, to America. In ancient Egypt, it was a rare luxury; today glass touches nearly every moment of our days. We drink our breakfast juice from it, cook in it, make lighting devices of it, keep the weather out with it, use it in television sets, automobiles, medical equipment, communication devices, and spaceships. It is difficult to imagine life without it.

In addition to familiar uses, glass has had its curious uses, too. Perhaps because it glittered with captured light, the ancients regarded it with awe and used it in magical amulets. The Chinese of early times customarily put glass cicadas,

imitating ones of jade, on the tongues of the dead because that insect was a symbol of life renewed after death. Napoleon lies under a shroud woven of glass fibers. Crystal balls, usually made of glass, are traditional foretellers of the future: steady gazing into their reflective depths may induce hallucinations which seem like distant events (crystallomancy is the name of this type of fortune telling). In the 18th century, colored and silvered hollow glass spheres called "witch balls" were hung in English cottages to ward off evil spirits. The families of fishermen often hung glass net-floats in the windows of their homes as a spiritual contact with the men at sea.

Whole orchestras have played upon glass instruments. Wigs with hair of spun glass have been made and a spun glass dress was fashioned for Queen Victoria. Victorian ladies carried glass hand coolers in the shape of eggs to cool their palms while dancing or being wooed.

Glass is important to many legends. Among the most persistent is that which relates how Alexander the Great, the 3rd century B.C. conqueror, went down under the sea in a huge glass jar to observe the plants and fishes there. Many Persian and Renaissance illustrations depict him in royal finery descending beneath the waves in his transparent bell, suspended by two ropes held by courtiers in a tiny boat on the water's surface.

In myths and fairy tales and also in scripture, glass is often a symbol of clarity, spiritual perfection, and revelation. Being solid yet transparent, it is used as a metaphor for a level of existence between the visible and the invisible, or between the mundane and the mysterious. Messengers from the Celtic otherworld, for example, sometimes arrive by sea in glass boats, symbolizing

the spiritual character of their mission. In Revelations 4:6, it is told that before the very throne of heaven "there was a sea of glass like unto crystal."

Glass mountains are often features of folk tales, to be climbed by the hero before he can win the princess. Or the true princess may be revealed by the test of the glass slipper. The folk hero often comes upon a glass palace after leaving the dark forest, as though coming upon the clarity of truth after the ordeal of ignorance. In the Celtic realm of the gods, a lofty glass castle manned by ghostly sentinels is the abode of the Perfect One. King Arthur's soul is housed in a glass castle in Avalon. Irish and Welsh legends tell of glass castles that are island shrines surrounded by glassy water.

These myths, marvels, and curiosities indicate the powerful fascination which glass has held for the human imagination. Samuel Johnson, writing on the subject in the *Rambler*, an 18th century publication, said: "Who, when he first saw the sand or ashes... melted into a metallic form... would have imagined that, in this shapeless lump, lay concealed so many conveniences of life...? Yet, by some such fortuitous liquefaction was mankind taught to procure a body... which might admit the light of the sun, and exclude the violence of the wind; which might extend the sight of the philosopher to new ranges of existence, and charm him, at one time, with the unbounded extent of material creation, and at another, with the endless subordination of animal life; and, what is of yet more importance, might... succour old age with subsidiary sight. Thus was the first artificer in Glass employed, though without his knowledge or expectation. He was facilitating and prolonging the enjoyment of light, enlarging the avenues of science, and conferring the highest and most lasting pleasures; he was enabling the student to contemplate nature, and the beauty to behold herself."

GLASS IN PRE-ROMAN TIMES

natural glass, obsidian, was formed over forty million years ago by volcanic eruptions. The intense heat fused masses of silica in a brown-black translucent glass, and from this hard material early man chipped tools and weapons.

Little is known, however, of man's first efforts to make glass himself. It is believed that glassmaking was invented—perhaps accidentally—in Mesopotamia (today's Iraq). Solid glass beads and amulets have been found which may have been made as long ago as 2500 B.C. The pendant shown here (1) was made 1000 years later. Later still, at the time of Tutankhamun (about 1350 B.C.), other glass jewelry which combined glass with gold and colored stones was made for royalty. The earliest hollow glass vessels (cups, bowls, bottles, etc.), more difficult to make than solid objects like beads, appeared in Mesopotamia and Egypt after 1500 B.C. The sphinx and the pyramids by then were already ancient monuments. It was two centuries before Moses would receive the Ten Commandments. Thutmose III (1490-1437 B.C.) established an Egyptian empire in the Middle East and traded with Phoenicia, Crete, and the Aegean Islands. It was probably after Thutmose had pushed into Mesopotamia and brought back glass that Egyptians learned how to make it.

Though the Egyptians were sophisticated enough to have an irrigation system, an alphabet, a canal connecting the Nile to the Red Sea, a knowledge of geometry, and contraceptives, the Egyptian glassmaker often used methods

1. **Star pendant. Northern Mesopotamia, 1450-1350 B.C.**

Opposite:

2. **Though many ancient glass vessels have been found, glass sculpture is extremely rare. This head of Amenhotep II, who preceded Tutankhamun by about 60 years as ruler of Egypt, is the earliest glass portrait known. It was cast in blue glass which, after its long burial, turned tan in color. Height 4 cm., 1436-1411 B.C.**

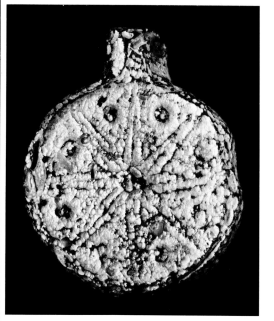

which seem primitive to us. To make glass containers, for example, he first made a core of clay combined with dung. Around the core he wound hot glass. After the vessel had cooled somewhat, he smoothed the still-warm glass with a flat tool or rolled it on a flat surface. He then trailed glass threads of brilliant colors—turquoise, blue, yellow, red, white—around the surface and dragged them up and down to form waves or feathery patterns. When the vessel was finished, he picked out the core with a pointed instrument.

In Egypt, no one but pharaohs, high priests, and nobles owned glass. Along with gold, silver, lapis lazuli, turquoise, ivory, alabaster, and carnelian, it enriched sumptuous thrones, golden funeral masks, regal sarcophagi, and magical protective jewelry. On the dressing tables of wealthy women stood glass containers for rare ointments, scents, cosmetics, and oils. The wig-wearing, opulence-loving aristocrats of Thebes and Alexandria placed a high value on glass vases, jugs, bowls, and cups for beer (the sale of beer was already regulated).

Elsewhere in the world of that time, Stonehenge was being built in Britain, the Chinese had learned to weave silk and to domesticate fowl and water buffalo, bathrooms were installed in the Minoan palace on Crete, and the Trojan War was underway.

3. **Core-formed amphoriskos. Egypt, 1400-1360 B.C.**

An Egyptian princess at her toilet. From *The Egyptians* by S.R.K. Glanville, A&C Black Ltd.: London.

5. Core-formed amphoriskos. Though produced many centuries later, it was made in the same fashion as earlier core-formed objects. Eastern Mediterranean, 2nd-1st c. B.C.

4. Core-formed alabastron. Mesopotamia, 800-700 B.C.

At roughly the same time, an independent glass industry was developing in Mesopotamia. Mesopotamian vessels were core formed in much the same manner as those made in Egypt.

Glassmaking declined from the 13th until the 9th century B.C., though it did not die out altogether. The industry was revived in Egypt, Mesopotamia, and elsewhere in the 9th century, at the time of King David and King Solomon. (It has been said that when the Queen of Sheba visited King Solomon, she saw in the courtyard of his palace what she believed to be a pool of water and accordingly tucked up her skirts so as not to wet them when she crossed, thereby revealing her regal limbs. Whereupon, according to Sir Thomas More, Solomon enlightened her by saying, "This is the palace evenly floored with glass.")

Core forming persisted as an important glass technique for many centuries, as demonstrated by the examples on pages 14 and 15. Some vessels were carved, as if they were made of semiprecious stone. Other glass containers were cast in molds, particularly in Mesopotamia. After casting, the surface was ground, probably by fast-running wheels fed with abrasives.

Core-formed glass was usually opaque. However, Mesopotamian cast glass was often a transparent pale green (6). Many cast-glass shapes were similar to forms used in metal. The cast bowl illustrated (7) has a raised, hollow area in the

7. **In Aristophanes's play, "The Archarnians" (1425 B.C.), an attack on the Peloponnesian War, the Greek ambassador to the Persian court relayed how "being guests we perforce drank the undiluted sweet wine from clear glass vessels and gold plate." Perhaps they drank from bowls not unlike this cast and cut, nearly colorless glass bowl. Probably Persia or Mesopotamia, probably 5th c. B.C.**

6. **Cast and cut vase of greenish glass with lug handles. The surface has been ground, probably by means of fast-turning wheels. Mesopotamia or Syria, 725-600 B.C.**

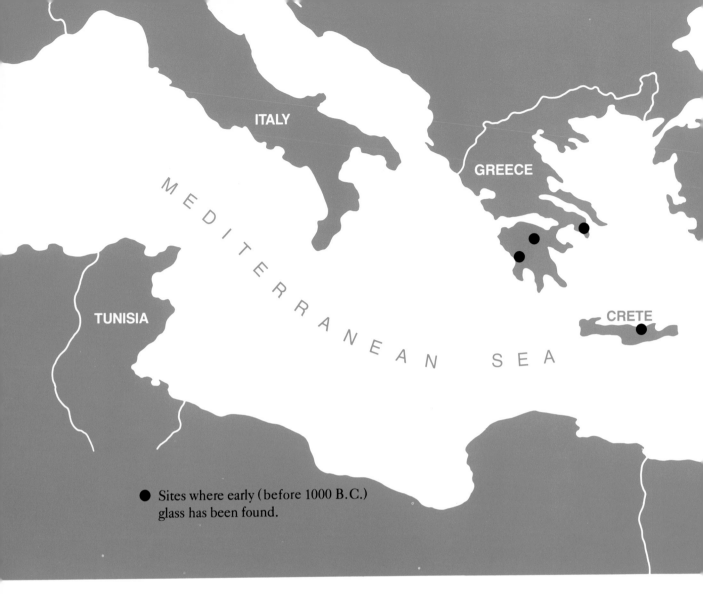

ITALY

GREECE

MEDITERRANEAN SEA

TUNISIA

CRETE

● Sites where early (before 1000 B.C.)
glass has been found.

8. **Hellenistic cut bowl.
Eastern Mediterranean,
3rd-2nd c. B.C.**

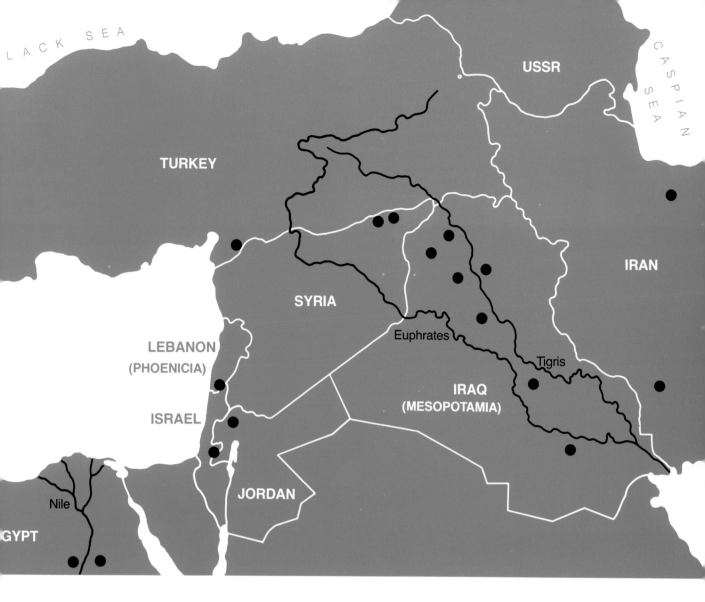

middle—an omphalos—to permit a secure grip: the holder put one finger inside the omphalos and his thumb firmly on the rim. To make this and other styles, finely powdered glass was probably reheated in a pottery mold until it all melted together. After cooling, it was ground and polished.

Middle Eastern and Egyptian glassmakers also made glass mosaics (9) from multicolored glass rods. The rods were fused together, forming a design, and the resulting large rod was heated and pulled out like taffy. The motif remained the same except that it got smaller and smaller as the rod was pulled longer and longer. The rod was then cut into slices of identical design, which could be arranged in various patterns or used individually as decorative inlays.

Egyptian and Mesopotamian glass has been found not only in the Mediterranean but as far afield as Russia on the one hand and France on the other, probably distributed by Phoenician traders.

9. **Mosaic glass plaque of an Apis bull, worshiped as a form of the supreme Egyptian deity, Osiris. Those who inhaled the bull's breath were thought to gain the gift of prophecy. The sacred bull was said to be the offspring of a virgin impregnated by a moonbeam. The Greek historian Herodotus reported that "Apis was a young black bull...on its forehead a white triangular spot." This spot can be seen on the plaque, which was probably inlaid on a piece of furniture. Egypt, 1st c. B.C. or A.D.**

GLASS OF THE ROMAN EMPIRE

By the birth of Jesus, glassmaking was nearly half as old as it is now. Yet the most significant event in this long history — the invention of glassblowing — had only just taken place. It probably first occurred about 50 B.C. somewhere along the Syrian-Palestinian coast, then part of the Roman Empire.

The invention of glassblowing was a major turning point in glass history. By blowing short puffs of air through a tube into a molten portion or "gather" of glass, rather than by casting or core forming it, a glassmaker could quickly inflate a bubble of glass and work it into a great variety (10) of sizes and shapes — or he might expand the bubble in a mold, giving it both form and decoration in a single operation. The Roman development of glassblowing marked the beginning of the mass production of glass.

From this moment on — by the use of sand, ashes, lime, fire, and his own life's breath — the glassmaker was able to produce an enormous and varied inventory. Glass was no longer exclusively a luxury product. In fact, it became more widely used for ordinary domestic purposes during the Roman Empire than at any subsequent time or place until the 19th century.

The Roman Empire at its height included what is now France, Spain, Portugal, England, Belgium, Switzerland, eastern Europe, Turkey, the Middle East, Egypt, North Africa, and parts of The Netherlands, Germany, and Austria. What is called Roman glass could have come from any of these places during the 1st to 4th centuries A.D.

The popularity of Roman glass rested not only on its usefulness and reasonable price, but also on its transparency and the beauty of its forms and colors. And, as Seneca (the Roman playwright and philosopher who was Nero's tutor) reasoned, the very fragility of the material further enhanced its popularity since "the desire for possessing things increases with the danger of losing them."

According to legend, a Roman glassmaker perfected unbreakable glass and presented a vase made of the material to the unpredictable Emperor Tiberius in his palace on the Isle of Capri. The craftsman displayed the beautiful transparent vase to the Emperor and then dashed it to the floor. It dented but did not break and the workman easily repaired the dent with his hammer as though the vase were metal. Tiberius asked if the workman had

Earliest known representation of glassblowing, found on a 1st century A. D. Roman lamp. Photograph courtesy of Split Archeological Museum. Yugoslavia.

Opposite:
10. **The invention of glassblowing made possible an enormous variety of vessels. Roman Empire, 1st-4th c. A.D.**

told the secret of unbreakable glass to anyone. The workman proudly assured him that he had not, whereupon Tiberius had him instantly beheaded, fearing the glassmaker's secret would destroy the value of all the Emperor's gold and silver. (Though some glass today is shatterproof, it cannot be worked with a hammer.)

Another witness of the Roman period, the poet Horace, said that despite glass's fragility he far preferred a transparent drinking glass to a metal cup: "Let my eyes taste, too, according to their capacity."

Drinking cups of all shapes and sizes were available both for household use and for the lively public bars and eating places which were found in most Roman towns. The cups were used primarily to serve wine, usually mixed with water, during and after the main meal of the day. Some cups were blown in molds with salutations such as "Rejoice and be merry" or "Cheers," toasts still heard 2000 years later whenever a glass is raised.

Mold-blown bottles came in fanciful shapes: animals, grotesque human heads, grape clusters. Mold-blown souvenir beakers might be embellished with chariot races. Others bore scenes of bloody gladiatorial contests (11) in which men fought to the death against each other or against wild beasts; thousands of animals might be slaughtered in one day to the delight of the crowd. The fans of these mass entertainments bought gladiator beakers as mementos of the event.

Some Roman glassmakers inscribed their names on their glass. "Ennion made me. Let the buyer remember him," is one of the most often-found inscriptions (13). Whether this was a testament of Ennion's pride in workmanship, or simply an advertisement reminding the buyer from whom to buy his next cup, no one knows — but it may be the first sales slogan.

11. **Gladiator beaker. 1st c. A.D. Gift of Arthur A. Houghton, Jr.**

The Colosseum in Rome, a 19th century view.

13. **Mold-blown ewer, inscribed under handle, "Ennion made me." Ennion was probably a Syrian glassmaker who migrated to Italy. Ennion's glasses are considered the most elegant of all the Roman mold-blown ware. 1st c. A.D.**

12. **Some Roman blown ware had a dappled effect, as in this vase. Before it was blown, it was covered with chips of colored glass and then blown to expand the patches of color. Roman Syria, 1st c. A.D.**

15. **Mosaic amphoris- kos. Egypt or Italy, late 3rd-2nd c. B.C.**

14. **Gold-band covered box, possibly for toilet- ries.**Constructed of bands of colored glass, some bands containing gold leaf. The bands were fused together and the vessel was then lathe turned and cut. **Ca. 50 B.C.-25 A.D.**

Romans particularly valued glass containers as shipping and storage vessels because they were transparent, reusable, odorless, and did not impart a taste. They were often decorated with an image of Mercury, god of commerce, and were packed in straw to survive long journeys by land and sea. Some were square for ease of packing.

Glass, in fact, was so widely used in Roman times that the Emperor Gallienus despised it as too common and would drink only from cups of gold.

Yet, at the same time that utilitarian glass was becoming commonplace, some of the most lavish glass ever made was produced. The opulent covered box shown here (14) has gold foil sealed inside its walls. Some elegant vessels were made of sinuous rainbow-colored bands fused together (16). Toiletry containers known as unguentaria were similarly made, some also containing gold leaf within the bands. These last may well have been among the most luxurious dressing table accessories ever made and emphasize the value placed on fine cosmetics and perfumes in Roman times.

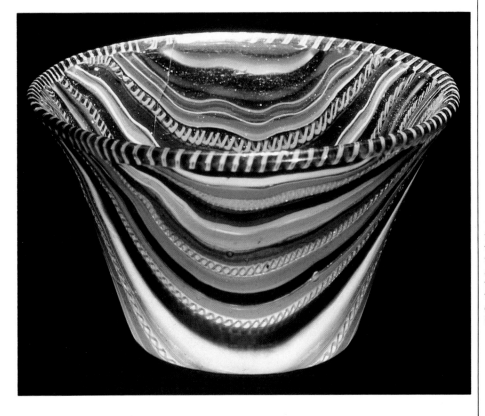

16. **Ribbon glass bowl, formed of canes twisted into cables. It was probably made in a number of stages, but the method of execution has been all but obliterated by polishing. 1st c. B.C.-1st c.**

Many Roman glassmakers sought to imitate rock crystal and other semiprecious materials. Layered stones such as those used for cameos—for example, onyx and agate—were emulated in glass (17) and carved in high relief to reveal contrasting colors. To make glass in this technique, the colored glass (often dark blue) was covered or "cased" with opaque white glass. It was essential that the two or more colored glasses be compatible in their rates of expansion and contraction; otherwise fatal cracks appeared. Once cooled, the piece was handed over to the cutters. The true refinement of their art is exemplified by the six-layered cameo (18), so finely cut that each layer was revealed, producing multicolored tonal and sculptural effects.

Techniques of enameling (19) and gilding on glass were also highly developed, methods not very different from those in use today. Colored

18. **Fragment of a six-layer cameo bowl. 1st c. A.D.**

17. **Cameo glass cup. Many cameos portrayed scenes of a mystical character. This one has been interpreted as a sacred ritual in which a Roman woman makes offerings to Priapus. Once in the collection of J. P. Morgan and sometimes referred to as the "Morgan Cup." 1st c. B.C. Gift of Arthur A. Houghton, Jr.**

26

19. **Enameled bottle. Depicts the mythological lyre-playing contest between Apollo and Marsyas. Late 3rd c. A.D.**

20. **Daphne Ewer.**
Opaque white glass
painted to depict the
myth of Daphne to
whom Cupid with his
dart had given a
passionate aversion to
the love-crazed Apollo.
In response to Daph-
ne's prayers, the gods
transformed her into a
laurel to save her just
as Apollo was about to
reach her. 2nd-3rd c. A.D.

glass, ground into a fine powder, was mixed with oil until it was the consistency of paint. It was then applied to the glass vessel and heated until the enamel fused permanently to the surface. The Daphne Ewer (20), on the other hand, is a famous example of cold-painted, non-fired Roman glass.

Other Roman vessels were decorated with both scratched and wheel abraded designs. The wheel abraded bottle here (21) was probably a travel souvenir from a seaside resort in the Bay of Naples.

The most beautiful glasses, some say, to be made in the Roman Rhineland were decorated with crimped strands of glass known as snake threads, similar to the

From a 6 A.D. Roman calendar, a figure representing August drinks from a transparent glass bowl. Reproduced from *Deutsches Archaeologisches Institut, 1888 Year Book*.

21. **Bottle, wheel abraded with a scene near Naples, possibly a souvenir. Inscribed in Latin, "Long life to you, happy soul." The scene is accurate enough to have helped modern archaeologists in their excavations at the site. 3rd-4th c. A.D.**

29

beaker from Cologne (22). The technique originated in the eastern part of the Roman Empire but artisans in the Rhineland brought it to its highest form.

The interiors of Roman villas were often ornamented with inlaid glass and cast glass panels imitating jasper, porphyry, and marble. According to Seneca, "A person finds himself poor and base unless his vaulted ceiling is covered with glass." Inlays of glass emulating precious stones may still be seen in the ruins of Nero's "Golden House" in Rome. Glass was also employed for lamps, hanging lights, and lanterns.

As the Roman Empire declined in the 5th century, its armies withdrew from Germany. Many local tribes, the Franks among them, came to dominate the area. Simpler glass styles resulted. A footless, horn-shaped drinking vessel like the cone beaker replaced the more elaborate Roman

22. **Snake-thread beaker. Probably from Cologne, 3rd c. A.D.**

23. Claw beaker. A characteristic Frankish glass, the claw beaker used hollow, clawlike or trunklike projections as decoration and to assure a good grip. 6th c. A.D.

24. Frankish cone beaker. 5th-6th c. A.D.

26. Octagonal pilgrim bottle. Both Jewish and Christian pilgrim bottles were sold to pilgrims in Jerusalem, differing only in their decorative symbols. According to the Talmud, the compendium of Jewish law and practice, legal action could be taken if glasses were broken by the crowing of a cock, the neighing of a horse, or the braying of an ass. Tyre, 6th-7th c.

styles. Its conical shape (24) reflected the lusty Frankish drinking habits: a glass was never set down upon the table but was drained at a gulp and taken from the drinker's hand to be refilled by host or servant—therefore these glasses did not need a base or foot to stand upon. (Footless glasses have persisted throughout glass history, down to the present day. Some have taken the form of hunting horns or are decorated with animal heads. Some Roman glass cups were modeled on the shape of a woman's breast, an adaptation of a style found in Greek pottery cups. A footless 17th century Dutch trick glass has a fanciful silver whistle on the end of the stem, possibly used to summon a refill.)

Other Frankish glasses had clawlike projections, such as that on page 31, or drops of hot glass applied which looked like green, red, purple, or brown "eyes," perhaps meant to ward off demons.

To the east of the Roman Empire, the Sasanians ruled Persia (now Iran) from 226 to 642 A.D. They were a cultivated, powerful people who enjoyed music, dancing, chess, and hunting, and their glassmakers developed a style of their own. It was distinguished by skillful carving in high relief (25) and the use of applied trail decoration. The Sasanians traded their glass throughout the Middle East, to Russia, and even to Japan where a number of examples have been preserved.

25. **Sasanian bowl with relief bosses. A great amount of time was required to cut away the surface between the bosses so that they stood in such pronounced relief. Glass of this kind was sufficiently precious to enter the treasuries of Byzantium, and from there be carried off by the crusaders. 6th-7th c. A.D.**

33

28. Red-cased vase, wheel cut through the ruby layer to the bubbly, colorless glass beneath. The ruby layer is worked three-dimensionally to represent galloping horsemen brandishing racquet-shaped objects; one holds a pair of lances. The neck is a temple with four men on a veranda. All these figures are shown in a craggy landscape wreathed in clouds. Reign of Emperor Ch'ien Lung, 1736-96. Gift of Benjamin D. Bernstein.

GLASS IN
THE FAR EAST

Eye-shaped beads, similar to Egyptian talismanic eye beads and dating from the Chou dynasty (1122-221 B.C.), have been found in China, suggesting a trade link between the two countries at an early date. During the Chou dynasty, one of the longest in Chinese history, both Confucius and Lao Tzu, the father of Taoist philosophy, lived and taught.

At roughly the same time as the Roman Empire and even earlier (221 B.C.-220 A.D.), the Chinese were producing small, jade-like engraved glass figures. In addition to these figures and the eye beads, pi disks (27) which symbolized heaven have been found to confirm the early, if limited, Chinese manufacture of glass.

A poet of the Chin dynasty (265-419 A.D.) tells us that Roman glass was imported to China, for he compares Roman glass to a spring day, and its clearness to the winter ice. Another early poet wrote of the difficult eastward

27. **Pi disk, symbolizing heaven. Probably Han dynasty, 206 B.C.-221 A.D.**

journey of Roman glass, recalling its dangerous trip across the scorching deserts and over the steep mountains. It is said that Chinese nobles and literary men of that time were so proud of their glassware that they invited guests for the express purpose of composing poems to celebrate their collections.

It is thought that in the 5th century, glassworkers from Persia may have come to China to engage in manufacture, and that at least one Chinese king, King Ch'en of Ho, sent a mission to Persia which brought back glass bowls among other treasures. However, all the evidence is vague until the end of the Ming dynasty (1386-1644) when Father Matteo Ricci recorded that he astonished the Chinese court by showing glittering Venetian prisms. "At the present time," he wrote, "they make glass, but inferior enough to our own." It seems likely that his Jesuit mission introduced glassblowing to the Chinese.

However, glass as a medium does not seem to have held much interest in China except as an imitation of jade, coral, and other materials. The inherent qualities of glass — its ductility and transparency — were generally ignored.

The preferred 18th century technique in China was cutting, often cameo, of brilliantly colored glass vessels, including ones of ruby red such as the masterpiece on page 34. Cut cased-glass snuff bottles (29) were in fashion, reflecting the highly ritualized practice of snuff-taking prevalent in China in the 18th and 19th centuries. (Snuff-taking is the inhalation of powdered tobacco.) Enameled imitations of fine 18th century Chinese porcelain were also made by Chinese glassmakers. The pair shown (30) were probably painted by artists at the imperial palace factory.

In India, as in China, glass was made to resemble jade, and imported glass was sometimes carved to resemble vessels made from stone.

29. **Five snuff bottles. 1800-1900.**

Opposite:

30. **Enameled Ch'ien Lung vases, made in imitation of porcelain. 18th c. Gift of Arthur A. Houghton, Jr.**

36

ISLAMIC GLASS

Roman influence on glassmaking diminished as the Roman Empire declined. Thereafter, European glassmaking in general produced only modest vessels for many centuries. During this period, called the Dark Ages in Europe, a new civilization arose in the area of glass's birthplace, the Middle East. Known as Islam, it developed between the 7th and 9th centuries and produced refined and elegant new styles of glassmaking.

The founder of Islam, of course, was the Prophet Muhammed, born in 570 A.D. in the city of Mecca in what is now Saudi Arabia. The word of Allah (Arabic for God) was revealed to Muhammed in a vision when he was forty. He became a preacher and in 622 established a community of his followers, called Muslims, in Medina, another Arabian city. He died in 632, and barely one hundred years later the Muslims had conquered a vast empire. It extended not only to the Sasanians in Persia but as far east as India and China; in the west it included Spain, Portugal, and Sicily. So enduring has been Islamic belief that today, though the empire no longer exists, Spain, Portugal, and Sicily are the only former conquests not still at least partly Muslim in religion.

By the 8th century, fabled Baghdad was the capital of Islam. Between 762 and 766, a hundred thousand workers built a new, circular Baghdad ringed by three walls of bricks, some bricks weighing 200 pounds. Baghdad's splendor-loving potentates encouraged an active cultural exchange — made possible by both conquest and extensive trade — between Baghdad and Europe and the Far East, further enhancing their opulent lifestyle.

The Muslims adopted something from every culture with which they came in contact. From the Chinese — who had reached such an advanced stage that their capital, Ch'ang, was the largest city in the world — the Arabs learned papermaking. They translated Euclid's *Elements of Geometry* from the Greek. From India, they adopted the numerals with which we all count. By 942, the Caliphate had learned about gunpowder from the Chinese and had established a postal system — and the tales of *The Thousand and One Nights* were being written. By the 13th century, Islam was more advanced than Europe in medicine, mathematics, alchemy, and optics.

Though the Koran, the Muslim holy book, denounced luxury, Muslims gloried in sumptuous interiors and richly embellished bowls, bottles, and other vessels, often made of glass which had been blown in molds, relief cut, or enameled. Perhaps this passion for lavish furnishings inside their palaces and holy places compensated for the outside environment of blistering desert and rugged steppe.

Glass imitative of rock crystal was immensely treasured by Islamic princes. Some of this glass was thought to be as valuable as gold and silver, substances prohibited by religion. The relief cutting found on these vessels was a difficult

Opposite:

31. **Relief-cut bowl.** Originally almost colorless, now weathered and iridescent, it is carved with birds, ibex, and decorative motifs. Its original appearance may be imagined if one recalls that the poet Abu Mansur Muhammed compared a glass vessel to ice and water in clarity. 9th c.

32. **Relief-cut bottle.** Two running goats confront each other on either side of a tree of life. 9th-10th c.

and costly process. It was achieved by outlining the design on the surface and then carefully cutting away the background to leave the design raised. This difficult technique is exemplified by the vessels illustrated on pages 38 and 39.

Glass imitations of turquoise were so convincingly made that they were sometimes mistaken for that precious stone. The story is told that in 1472 the Shah of Persia presented a turquoise bowl, already centuries old, to Venice. On its base was inscribed "Khorasan," a province famous for turquoise mines. Neither the giver nor the recipient realized that the bowl (which may still be seen in the treasury of Venice's St. Mark's Cathedral) was made of glass.

Egypt, having now become part of the Muslim world, contributed importantly to the art of enameled glass by the refinement of a luster stain made with silver. After the silver stain was painted on the glass, the object was heated to produce brown and yellow colors.

Some later Islamic relief-cut vessels are known as Hedwig beakers (33), named after a European Christian saint despite the fact that the glasses are Near Eastern in origin. St. Hedwig is said to have used a glass in this style. She would drink nothing but water from it — yet when her husband tasted from the glass, the contents had miraculously turned to wine. A number of glasses associated with the saint, who died in 1243, came to be treated as holy Christian relics. One, belonging to St. Hedwig's sister, St. Elizabeth, later came into the possession of Martin Luther and was credited with miraculous powers at childbirth.

33. **Hedwig beaker, once used as a gluepot in the sacristy of Halberstadt Cathedral in Germany. Only fourteen of this type of beaker are known to survive today. 11th-12th c.**

Images of living things were forbidden by the Islamic religion; however, as with gold, the prohibition was not always followed, especially when making objects for private use. Islamic glassmakers often decorated their wares with rhythmically repeated patterns of plants (34), geometric designs, and quotations from the Koran. If Muslim glassmakers and other craftsmen represented human beings or animals, they did so in highly stylized form. Fishes were a favorite decorative theme on enameled glasses; emblems of good luck, they may have provided magical assurance against poison.

Of life in the great Islamic glass center of Damascus, a 14th century visitor wrote home: "Concerning the wealth of this city...which shows forth particularly in gold and silver, cloth of gold and silk, in gold [despite the Koran], silver, and bronze vessels incomparably fashioned with great art in the Saracenic manner, in glasses most pleasingly decorated which are commonly made in Damascus, we forbear to write, for they cannot be captured on paper nor set forth in words." The glasses he was referring to were probably enameled and gilt vases (35), perfume sprinklers, beakers, and bottles made first in Aleppo and later in Damascus which reached a height of artistry not equaled again until Renaissance Europe. One such enameled beaker is the subject of an English legend: The beaker, called the Luck of Edenhall, was probably brought to England by a crusader, but the story says it was left by fairies near a spring at Edenhall and that the luck of the residing family was dependent on its safekeeping. When one of the

34. **Mold-blown bottle with frieze of plants above a Kufic inscription. This bottle also has a band of applied threads. Many Islamic vessels had long thin necks which kept the contents from evaporating. 11th-12th c.**

Detail of a 16th century Persian miniature, showing a bottle similar to the bottle above. From the Houghton *Shah-nameh (Book of Kings).*

41

family took it, the fairy king warned, "When this cup shall break or fall/ Farewell to the luck of Edenhall." The beaker now resides at the Victoria and Albert Museum in London—intact, as is Edenhall.

One of the most lucrative branches of the Syrian glass industry supplied intricately enameled lamps (36) to mosques, the Islamic houses of worship. It has been said that in Istanbul's Hagia Sophia—originally built as a Christian cathedral in the 6th century, becoming a Muslim temple or mosque after 1453, and now used as a museum—the blazing multitude of hanging lamps gave the impression of a ceiling of light. Mosque lamps housed a separate holder for a bowl of oil on which floated a wick. These lamps not only lit and decorated the mosques, but were symbolic of the light of God as well.

The beauty of Islamic glass was marvelously conveyed by an 11th-12th century Islamic writer who said:

> *He set before us whatever is sweet in the mouth or fair to the eye....He brought forth a vase, which was as though it had been congealed of air, or condensed of sunbeam motes, or molded of the light of the open plain, or peeled from the white pearl.*

The manufacture of Islamic glass ended when the Mongol conqueror Tamerlaine destroyed Damascus in 1400, carrying the glassmakers off to his capital of Samarkand.

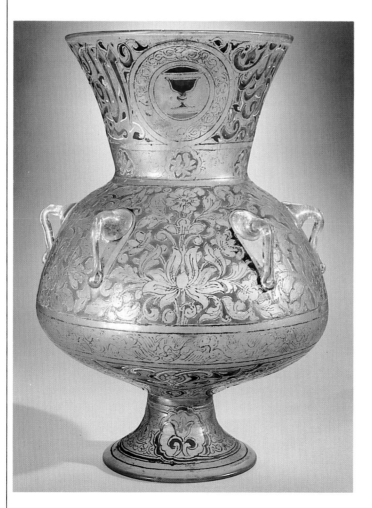

36. **Enameled and gilt mosque lamp. It is inscribed, "God is the light of the heavens and the earth. His light is as a niche in which is a lamp, the lamp in a glass, the glass as it were a glittering star." Syria, ca. 1355.**

35. **Enameled and gilt vase, decorated with fish, arabesques, lotus heads, rosettes, and a repeated inscription, "The Wise." The lotus flowers are a clear sign of Chinese influence as the lotus was a typical Chinese theme. Damascus, 1320-1330.**

EUROPEAN GLASS

We have seen how advanced not only glassmaking and other crafts were during the Middle Ages in the Islamic Empire, but how developed was Islamic learning. Other cultures, too, had reached a high level. During the centuries called "dark" in Europe, the Chinese were already printing books, in Japan the Baroness Murasaki was writing *The Tales of Genji*, and in Persia Omar Kayyám was composing his timeless love poems.

However, for most of Europe's population in the Middle Ages, life was difficult, even precarious. People lived in small, isolated villages or on the land. Famine and plagues occurred frequently. Many were endangered by Vikings who captured people in central Europe to sell as slaves in the Middle East (the Vikings were also discovering Iceland and Greenland and later, some say, North America). What learning and skills remained were preserved in the monasteries. European glassmaking was at a low ebb. Only a few primitive vessels were being produced.

By the 12th century, with the growing power of the Catholic Church, the darkness in Europe began to lift. The development of Gothic architecture in northern Europe stimulated the production of brilliantly colored glass for cathedral windows at Chartres, Canterbury, and elsewhere. While it was artists and glaziers who made the elaborate stained glass windows, the actual colored glass used by them was produced by glasshouses, many of them connected to monasteries.

Most stained glass windows illustrated Bible stories and helped convey Christianity to people who could not read, which was most of the populace, including the lords and ladies. "Bright is the noble edifice," said Abbot Suger, speaking both literally and figuratively of the effect of these windows, "that is pervaded by new light." To the abbot and others, the beautifully tinted light, streaming into the new cathedrals, not only illuminated the tales of the Bible but symbolized the truth of God.

In the south, where people were as unlettered as in the north, gloriously colored religious mosaics adorned churches in Greece, Italy, Sicily, and Byzantium. They were composed of colored glass cast as thin, flat cakes and cut into small pieces. By the fall of Byzantium in the 15th century, millions of these little pieces of glass must have been made to form rich, golden representations of episodes from the Old and New Testaments.

It was probably the manufacture of glass for church windows which kept the technique of glassmaking alive during the Middle Ages, since the manufacture of glass vessels remained so crude. However, little by little, more refined utilitarian glass was again produced in both southern and northern Europe.

VENICE

The first important glass center in southern Europe was Venice, the queen of the seas whose ruler, the Doge, married the Adriatic every year in a rite during which, to the accompaniment of hymns, prayers, and guns, he tossed a diamond ring into the water — thus emphasizing Venice's position as the major sea power and trade center of the time.

As the Islamic Empire declined, Venice became the crossroads of East-West trade; she had a monopoly on commercial shipping to the East from the 11th century. It was probably in the course of peaceful trade with the Middle East

Sixteenth century view of the Piazzetta S. Marco, Venice, by G.A. Vavassore, after Jost Amman. Courtesy of the Metropolitan Museum of Art, New York.

that Venetians received glass gifts and learned the art of glassmaking. In addition, returning crusaders brought many fine specimens of Islamic glass to the treasury of Venice's great cathedral.

So opulent was the shimmering, domed, and arcaded city of Venice that any treasure could be bought or sold there: spices, scents, sables, slaves, gold, silver, jewels, ivory, apes, ebony, silks, lace, damasks, velvets — and glass. "All the gold in Christendom," according to one medieval chronicler, "passes through the hands of the Venetians." These wealthy, flamboyant men not only indulged themselves with exquisite glass, among other luxuries, but also marketed it to the world. By 1317, at the very beginning of the Renaissance, Venetian trading ships laden with glass had already landed in Flanders, opening the whole of northern Europe to the influence of Venetian styles.

The Venetian glassmakers' guild was formed in the early 1200s. In 1291, the glass industry was forced by the authorities to move from the city to the nearby island of Murano — where it has remained to this day — so that both the danger of fire from the furnaces in the city itself might be eliminated and the glassmakers more effectively controlled. It was not such a hardship, since Murano was a summer resort where Venetian aristocrats had villas. Only an hour's row from Venice, the lagoon between was studded on warm nights with gondolas going to and fro.

Left to right:

37. *Cristallo* goblet, decorated with enamel and gilt cherubs riding on garlands. Venice, early 16th c.

38. Dragon-stemmed goblet. The dragon motif, used in Venetian and Venetian-style glassware, may have come from the story of St. George and the Dragon, or may have had its source in the Far East where the dragon was sacred. Venice, 17th c.

39. Covered goblet with striped decoration. Twisted cables of white glass alternate with plain white stripes. The cables and stripes were embedded in a gather of clear glass, then pinched and twisted, coated with another layer of *cristallo*, and blown. Venice or northern Europe, late 16th or early 17th c.

40. Wineglass, composed of three separate bubbles of glass. Glasses like this are seen in late 16th century paintings. Such pure form expresses the inherent qualities of the glass and does not depend on decoration or complication. Venice, mid-16th c.

Though Venetian control over the glassmakers and their families was strict, these craftsmen were highly regarded. Many were given patrician standing, their daughters being permitted to marry noblemen.

Control was crucial to Venetian trade because the glassmakers were privy to many jealously guarded secrets regarding the construction of furnaces, the formulas and proportions of ingredients, and the making and handling of tools. Glassmakers at that time relied on trial and error, using their eye, their judgment, their past experience, and the knowledge handed down from glassmakers before them. Venetian glassmakers were not allowed to leave Murano; in fact, escape was a crime which carried the death penalty. There are stories of relentless assassins following fleeing glassmakers—in one case, to the very gates of Prague. Yet many glassmakers somehow did leave, emigrating and setting up manufactories in the Tyrol, Vienna, Flanders, Holland, France, and England. It was not until the 17th century that a book written specifically to instruct glassmakers was published by a Florentine, Antonio Neri. His book, *L'Arte Vetraria* (The Art of Glass), printed in 1612, made available to everyone secrets which had long been carefully guarded.

So important to Venice was the glassmaking craft that when Henry III of France visited Venice in 1574, it is said his ship was greeted by glassblowers, borne on a huge raft, who blew objects for the king's amusement, their furnace in the shape of a mammoth sea monster spewing flames. (The king was then entertained at a vast banquet featuring a menu of 1200 dishes.)

Venetians apparently made glass beads from the 13th century onward. These beads were used in rosaries and jewelry and were an important part of the African slave trade. Following the discovery of America, they became extremely

View of Murano, 1559. Published by Matteo Pagan.

popular with North American Indians. According to Christopher Columbus's log for October 12, 1492:

> *Soon after, a large crowd of natives congregated. . . . In order to win the friendship and affection of that people. . . I presented some of them with red caps and some strings of glass beads which they placed around their necks, and with other trifles of insignificant worth that delighted them and by which we have got a wonderful hold on their affections.*

From the 15th to the late 17th century, Venetian glass mirrors were considered the finest in the world. The glass had to be very even and clear to reflect properly. It was backed with metal foil. Before this, mirrors had been made of steel or polished silver.

By the middle of the 15th century, and perhaps even earlier, Muranese glassmakers produced an almost colorless, highly esteemed glass called *cristallo* (37, 40), an allusion to rock crystal. They also made a wide range of colored glass: dark blue, amethyst, red-brown, emerald green, and milky white. In addition, they made various kinds of marbleized, millefiori (meaning "thousands of flowers"), lace, and aventurine (literally, "adventure") glass, the latter appearing to have gold dust sprinkled within it. A 16th century visitor to Murano, the Anglo-Welsh writer James Howell, had a fanciful notion: he thought the superiority of the island's glass was due to "the quality of the circumambient Air that hangs o'er the place." However that may be, it was the Venetian glassmakers' skill that made their glass supreme, as the examples illustrated on pages 46 and 47 attest. Their elaborate vessels, often worked in complex, virtuoso shapes (38, 39), were sought as the height of grandeur by all who loved to make a splendid display of wealth.

42. **Spanish** *cantir.* **Spanish glass, while employing Venetian techniques such as the filigree handle on this example, also showed the strong influence of the Muslims who controlled Spain. The shape of the** *cantir* **– a wine-drinking vessel with a long thin spout – is, however, specific to Spain. To drink from it, one holds it at a distance and tilts it to spurt a stream of liquid directly into the mouth. It was devised to make communal drinking convenient without anyone's lips touching the vessel – yet another example of the drinking customs enlivening glass history. 18th c.**

41. **Covered** *cristallo* **beaker. It represents a purer, less elaborate Venetian style sometimes used for** *cristallo.* **Such large cups served as communal drinking vessels at meetings of German guilds from the Middle Ages to the 19th century. This one is inscribed with diamond-scratched signatures of innumerable members of some such society. The signatures date from 1574 to the late 18th century. Venice or northern Europe, ca. 1530.**

Some gullible people believed that the best Venetian drinking glasses would shatter instantly should they come in contact with the least drop of poison—more a sign, perhaps, of the conspiratorial character of the times than the nature of the material. In any event, Venetian glass was indeed very fragile; merchants complained constantly of excessive breakage in shipment. Commenting on this fragility, a 16th century writer, Vannoccio Biringuccio, said:

> *Considering its brief and short life owing to its brittleness, it cannot and must not be given too much love, and it must be used and kept in mind as an example of the life of man and of the things of this world which, though beautiful, are transitory and frail.*

Opaque enameled decoration on both colored glass and *cristallo* frequently portrayed classical themes. Others, such as the bowl pictured here (43), were in the style of the painting of the time. Sometimes these scenes were shown on a field of gold which had been fused to the surface in the furnace.

Diamond point engraving was another popular means of glass decoration, often consisting of simple, elegant floral motifs. Diamonds were just entering Europe and, India being then the sole source, they probably all came through Venice—so diamond point engraving was a natural Venetian development.

With the progress of science during the Renaissance, glass was found to be ideal for laboratory apparatus. It could be fashioned into complicated forms and it neither corroded nor interfered with most chemical or pharmacological procedures. Said one Jeremias Martius in 1584, "Man uses glass for many things, but its service in medicine surpasses all others."

The optical properties of glass permitted improved eyeglasses and the invention of microscopes and telescopes. (Without Galileo's observations through the latter, he might not have conceived new theories about the cosmos which resulted in his imprisonment.)

While Venice was the dominant center of glassmaking in southern Europe, glassmaking is known to have been practiced since the 12th century in the Italian village of Altare, situated in the mountain pass where the Alps meet the Apennines. Little is known of Altare glass, but it was probably similar to the Venetian in style. Many of the Altare glassmakers migrated, some settling in France. Unlike the prohibition against the migration of Venetian glassmakers, the glassmakers of Altare were actually encouraged by the authorities to go abroad and work.

NORTHERN EUROPE

North of the Alps, anyone with pretensions to a refined life owned, used, and displayed expensive Venetian glass. The insatiable desire for this glass, its cost, and the difficulty of getting it caused many to dream of manufacturing it. Eventually, hundreds of Italian glassmakers—despite the ban on emigration—were enticed to northern cities with promises of wealth and social prestige.

Although the influence of Venice (44) eventually dominated luxury glassmaking in Spain, France, Germany, The Netherlands, England, and even Sweden in the 16th and 17th centuries, other kinds of glass were also being made in northern Europe, principally in the forest glasshouses.

Opposite:
43. *Lattimo* bowl with enameled and gilt decoration. Venice, ca. 1500.

44. Two goblets with elaborate stems which show how the Venetian influence developed in the north. The dragon-stemmed goblet was made in The Netherlands and the covered goblet was made in Stockholm.

Some forest glasshouses were large, long-lasting settlements complete with cottages for workers and a manor house for the owner. Others were small, sometimes short-lived enterprises near a monastery or the estate of a nobleman. In both cases, their systematic clearing of the forest for fuel helped to reduce the isolation of local villages. More importantly, at a time when famine beset the people regularly, their efforts cleared land for agricultural use. Consequently, the impact of forest glasshouses often transcended glassmaking alone.

The forest glasshouses were run by families and guilds who, like the Venetians, kept their secrets to themselves. To become a glassmaker was difficult because the guilds had uncompromising rules. "Nobody shall teach glassmaking," said one, "to anyone whose father has not known glassmaking.... Nobody shall learn unless his father has promised and sworn and belongs to the guild and has made glass. The art may be practiced only by male children of legitimate marriages within specific glassmaking families." Boys entered the craft at the age of twelve and had to swear never to show "the said noble art, usage, and science" to anyone outside the guild.

Glassmaking was a rigorous, hazardous occupation. According to a 1713 book:

> *Certainly no one could endure for long the strain of such work as these men have to do, nor could it be kept up except by robust men in the prime of life. During the process of making glass vessels the men stand continually half-naked in freezing winter weather near very hot furnaces and keep their eyes fixed on the fire and molten glass ...their eyes have to meet the full force of the fire...they shrivel because their nature and substance...is burnt up and destroyed by the excessive heat.*

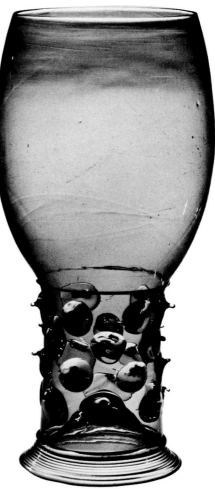

45. *Waldglas* beaker with looped prunts. Called *waldglas* because such glass was made in forest glasshouses. *Wald* means "forest." Germany, 16th c.

46. *Roemer.* Sometimes these vessels were enameled or engraved by diamond point but the design of this one depends on its own handsome shape without benefit of decoration other than typical knobs on its hollow stem. The Netherlands, 1600-1625.

When undertaking the risky, tricky task of changing a cracked or worn-out melting pot while the furnace was roaring with flames, the forest glassmaker dressed in garments and mask or hood of the skins of wild animals as protection. He was often black with soot from the fire and his bizarre costume made him so terrifying a sight that he was considered by some parents as a monster with whom to threaten misbehaving children.

The ingredients of northern forest glass were different from glass made in the south. The sand contained iron, which produced forest glass's typical green. Rather than import ingredients from the Mediterranean region, the northern forest glassmaker used the wood ashes (potash) which were the natural by-product of his own wood-burning furnace. By carefully purifying the ashes and adding copper oxide, he deliberately nurtured his glossy green. Even after he learned to make colorless glass in the 16th century, he and his customers often preferred the typical green of his wares, enjoying the pleasant tint it gave the wine. Today, Rhine wine is traditionally served in green glasses.

The most usual products of the northern forest glasshouse were window glass and large drinking vessels. Some of these beakers were of immense size, attesting to the notorious capacity of the users. The beakers had knobs on the surface to allow a firm grip (45); the drinker's hands were apt to be greasy inasmuch as forks and napkins had not come into popular use (although affluent Venetians had eaten with forks since the 11th century). In 1564, a German minister, referring to glassmaking in a sermon, spoke of these knobbed beakers:

> *Nowadays one applies buttons, prunts, and rings to glasses to make them sturdier. Thus, they can be held more easily in the hands of drunken and clumsy people.*

A more refined drinking glass (46), called a *roemer,* had a flared or rounded

The extraordinary size of so many German drinking vessels inspired one writer in 1688 to say: "You know the *Germans* are strange Drinkers....Every Draught must be a Health, and as soon as you have emptied your Glass, you must present it full to him whose Health you drank. You must never refuse the Glass which is presented, but drink it off to the last Drop. Do but reflect a little on these Customs, and see how it is impossible to leave off drinking....To drink in Germany is to drink eternally...."

Le Buveur Flamand (Flemish Drinker). Engraving by P. Chenu after a painting by David Teniers, France, 1743.

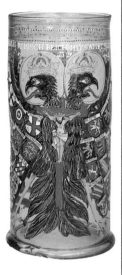

bowl; the knobs were confined to the hollow stem where they still afforded a secure grip. *Roemers* are seen in many paintings of the time depicting both homely peasant life and high society.

Large cylindrical glasses, called *humpen,* held several quarts of beer and were among the most popular enameled glassware. Martin Luther called them "fools' glasses" and "obscenely large welcome cups," alluding to the custom of draining one in a single gulp.

German and Bohemian glassmakers enameled drinking glasses with patriotic designs, biblical and mythological subjects, coats of arms, and scenes of daily life. Some depicted political themes such as the unity of the Holy Roman Empire (47) or the ending of the Thirty Years' War.

Many styles of engraved glassware produced in Bohemian factories and elsewhere became increasingly popular in the northern markets. Delicate diamond point engraving (48) rivaled enameling as a desirable decoration. Diamond point engraving required no prolonged training, the most important requirement being the ability to draw well. This made it possible for amateurs to pursue it as a hobby, often with calligraphy their main interest. (Sixteenth century Elizabethan poets in England sometimes engraved their verses on window panes in diamond point calligraphy.) As glassmaking developed in the north, and tastes changed, northern-made glass finally replaced Venetian glass as the most fashionable (49).

49. **The Maximilian
Pokal.** Goblet engraved
and signed by Johann
Wolfgang Schmidt, one
of the greatest engrav-
ers who ever lived. It
depicts Maximilian II
Emanuel, Elector of
Bavaria. On the reverse
side is a battle scene.
Such goblets were ele-
gant status symbols.
Nuremberg, Germany,
1690. Bequest of
Jerome Strauss.

The development in Bohemia of a formula for a brilliant, colorless, easy-to-cut glass facilitated the perfection of wheel engraving. This formula used chalk (limestone) as a principal ingredient and its use spread quickly to other glass centers on the Continent. The hardstone carvers of Prague, Munich, Nuremberg, and elsewhere turned from carving rock crystal and gemstones to engraving this new glass. They would press the glass against rapidly rotating wheels of copper or stone to incise designs and scenes. As in enameling, subject matter included coats of arms, portraits of princes, the ages of man, the seasons, mythology, and courtly life. The vessels were sculpturally cut in relief (raised above the surface) or engraved in intaglio (cut into the surface)—and often in a richly wrought combination of both (50) which has rarely been equaled since.

Bohemian glasshouses sent their salesmen throughout the world to seek new markets. These vendors were often trained cutters and engravers themselves who could decorate glassware on the spot to the individual wishes of their customers.

50. **Covered chalk glass beaker, richly engraved in both intaglio and relief. Germany, Potsdam, 1690-1700. Gift of Edwin J. Beinecke.**

51. **Wheel engraved plaque. Engraved by Hans Wessler, a goldsmith said to have introduced glass engraving to Nuremberg, the plaque shows Tomyris, Queen of the Messagetae, cutting off the head of the dead Cyrus, King of the Persians, who had invaded her kingdom and killed her son. Nuremberg, 1610-1620.**

ENGLAND

Though glasses of a rude sort and fine, colored window glass had probably been made in England since the 13th century, sophisticated glassmaking seems to have been initiated there, as in so many other European countries, by Venetian artisans. In 1571, Giacomo Verzelini came to London from Antwerp where he had been working. With nine other Italian glassmakers, he joined an English glasshouse and made glass in the Venetian fashion. For about one hundred years thereafter, the Venetian style dominated English glassmaking.

The Venetians were working in London during the Elizabethan Age, the time of William Shakespeare, Ben Jonson, John Donne, and Christopher Marlowe, of Sir Walter Raleigh and Sir Francis Drake. England was in a period of expanding exploration, trade, and colonization; she was also faced with a severe shortage of wood needed for the ships engaged in these pursuits. By 1615, there was a prohibition against using wood for fuel; glassmakers, switching to coal, had to overcome many new problems, such as how to prevent coal fumes from ruining the molten glass.

English glassmaking increased substantially with the mid-17th century English development of a black (actually dark green) bottle which protected its contents from light. Thick-walled, it was also very durable and few broke in shipment. With the advent of the black bottle, England became the foremost supplier of bottles to the western world for more than a century. Daniel Defoe, in *Robinson*

52. **Ravenscroft lead glass *roemer*, marked with his raven's head seal. Decorated with mold-blown ribbing which, on the bowl, is pinched to form a mesh design. England, 1676-1677.**

An 18th century English glasshouse. From a 1747 engraving by C. Grignion.

Crusoe (1719), wrote that "great Numbers of Bottles…are now used sending the Waters of St. Vincent's Rock [an English spa] away, which are now carry'd, not only all over England only, but, we may say, all over the world."

A clear, colorless glass less breakable than Venetian *cristallo* had long been sought in England. In 1676, working in London in great secrecy, an English glassmaker who had lived many years in Venice, perfected a formula for lead glass. His name was George Ravenscroft and his brilliant, heavy, resonant material is as highly regarded today as when he first developed it. Ravenscroft marked his new glasses with a raven's head seal, derived from his coat of arms. (One of his glasses is shown on page 57.)

Other English glassmakers soon took up Ravenscroft's lead glass. It was an essentially brand new material, never blown or worked by glassmakers before and quite different to work than Venetian *cristallo*. It remained in a workable condition longer; that, and its weight, clarity, and capacity to capture light led to a new goblet style: simple, undecorated shapes which relied on graceful form and fine proportions. A variety of new stem formations—using knops, balusters, teardrops, and air twists—provided a play of deep shade and brilliant light such as had never before been seen in glass.

53. **Williamite political glass inscribed in opposition to the Jacobite cause: "To the glorious pious and immortal memory of the great and good King William who freed us from Pope and popery, knavery and slavery, brass money and wooden shoes and he who refuses this toast may be damned crammed and rammed down the great gun of Athlone." Ca. 1740-1750. Bequest of Jerome Strauss.**

54. **Diamond stippled goblet. Engraved with a portrait by the eminent Frans Greenwood in The Netherlands on English lead glass. 1746.**

Engraved glasses carrying political messages were popular after the bloodless revolution which ended Roman Catholic rule in England and brought William III of Orange to the throne in 1689. Political sentiments were strong among both William's supporters (53) and his opponents, the Jacobites, who advocated the return of the Catholic King James or his sons. An inscription on one Jacobite glass reads:

God Bliss the Subjects all/And Save both Great and Small
In every Station./That will bring home The King/Who has best Right to Reign,
It is the only Thing/Can Save the Nation.

Jacobites were liable to charges of treason for possessing such a glass.

Much English glass was exported to the Continent, principally to The Netherlands, where it was wheel or diamond point engraved. The most distinguished diamond point engraver of his time was Frans Greenwood. He developed stippling, a method of composing a scene or portrait (54) by making dots rather than drawing lines with a diamond pointed tool. Where the dots were close together, engraved areas appeared as light, and the untouched areas appeared dark. Diamond point had a delicate tonal appearance, as compared to wheel engraving which was usually more sculptural.

55. **Enameled and gilt goblets. Two members (brother and sister) of a family of pre-eminent English enamelers named Beilby, working in Newcastle-on-Tyne, practiced the technique of enameling clear crystal. These two goblets bear the arms, crest, and motto of the Earls of Pembroke and Montgomery and are signed. 1765-1770.**

Glass cutting was introduced into England by German and Bohemian craftsmen, but a distinctly English style quickly emerged. The surface was covered with an orderly geometric pattern of facets which brought out the high refractive index of the lead crystal, causing it to sparkle brilliantly. This style was used on drinking glasses, centerpieces, candlesticks, and for the newly developed prismatic chandeliers. It must be remembered that 18th century rooms, even in richly appointed houses, were very dark, depending for illumination on candles, which were expensive and, in England, heavily taxed. It is said that in 1772, Queen Charlotte's large dressing room, hung with crimson damask, was lit by just four candles. Only on festive occasions were homes brightly lit—as when in 1769 Lady Cowper wrote proudly that she had "an assembly in my great room, with about five dozen wax lights in the room." So, as cut lead-glass candelabra (56) and chandeliers became prevalent, their light-reflecting qualities multiplied the candlelight, enlivening many an elegant drawing room.

A heavy excise tax had been levied on English glass, but not on Irish glass. When Parliament lifted a thirty-five-year ban on exportation of Irish glass in 1780, the tax-free Irish glass industry began to turn out enormous quantities of glass for the large American market. Ireland became financially more advantageous than England for the glass industry and much skilled English labor moved there. As a result, styles in English and Irish cut glass were similar, and it is difficult to tell them apart (57). Waterford, a name that has become synonymous with fine Irish glass, was the city in which one of the glasshouses

56. **Candelabrum. England or Ireland, ca. 1785.**

60

operated. An advertising circular from a glass store in Cork about 1790 gives an idea of the range of Waterford glass:

> …*His shop is now completely stored/With choicest glass from Waterford–/ Decanters, Rummers, Drams and Masons/Flutes, Hob Nobs, Crofts and Finger Basons, Proof bottles, Goblets, Cans and Wines, Punch Juggs, Liquers and Gardevins;/Salts, Mustards, Salads, Butter Keelers/And all that's sold by other dealers,/Engraved or cut in newest taste,/Or plain–whichever pleases best;/Lustres, repaired or polished bright,/And broken glasses matched at sight;/Hall globes of every size and shape,/Or old ones hung and mounted cheap.*

However, Waterford was only one of many cities that boasted a glassworks. Others were in Dublin, Belfast, Cork, and elsewhere. They flourished until the mid-19th century, and in the mid-20th century some of them were revived.

Although European glass had been exported to some extent to the Far East in the 17th century, it was the ascendancy of the English East India Company in the early 18th century which accounted for the great increase in English exports to India, exports sold both to English residents and the native population. Among these exports was glass. For example, it is recorded that shortly after 1721, five chests of glassware together with a telescope and thirteen looking glasses formed part of an outward consignment valued at £500, a considerable sum at that time. A 1737 advertisement mentions the export of over 6000 pieces of glass. This glass trade with India, combined with the American market for English and Irish glass, provided a lively overseas market for the British.

The glass showroom
of Pellatt and Green.
London 1809.

57. **Cut glass fruit bowl.
Ireland, ca. 1790.**

AMERICAN GLASS

Glassmaking was America's first industry. When the Jamestown colony was established in 1607, the settlers brought glassblowers with them. However, the efforts at Jamestown to manufacture glass—and other attempts to follow near Philadelphia and Boston—failed despite the abundance of fuel and good sand. Throughout the 17th and 18th centuries, the colonists imported most of their glass.

In 1739, more than a hundred years after Jamestown, Caspar Wistar established a factory in southern New Jersey. Despite a British ban on manufacturing in the colonies, Wistar, a German, imported German glassmakers to produce window glass, bottles, and tableware (58). Benjamin Franklin was an early patron of Wistar, and used Wistar's scientific glassware in his electrical experiments. Caspar Wistar, the first commercially successful glass manufacturer in this country, began the German domination of the American glass industry that continued until the 19th century.

The second German to manufacture glass successfully in the New World, Henry W. Stiegel, built three glasshouses between 1763 and 1774 in Manheim, Pennsylvania. Stiegel, too, imported workmen, both from Germany and from England, and sometimes he lured glassmakers away from Wistar. The demand for all kinds of glass in the colonies was huge and, though most of the colonists' glassware came from England, Stiegel sold enough to prosper handsomely. According to legend, he styled himself a baron and lived in a manner befitting nobility, even employing musicians to strike up a tune whenever he arrived

Opposite:

58. Green covered sugar bowl, made in the *waldglas* tradition. Perhaps from Caspar Wistar's glasshouse in southern New Jersey, ca. 1750-1777.

59. Blue sugar bowl. Possibly Manheim, Pennsylvania, glassworks of Henry W. Stiegel, ca. 1765-1774.

home in his coach. He was an active churchman and deeded land to the Lutheran church in Manheim, asking in return only the annual payment of one red rose—forever. The flowery rent is still sentimentally paid by the church the second week of every June to a descendant of Henry Stiegel.

Both Stiegel and Wistar eventually failed, Stiegel because he overextended himself financially, and Wistar probably because of the disruption of the Revolution.

A third German, John Frederick Amelung, opened a large glass factory in Maryland in 1784, the year the War of Independence ended (it was also the year Benjamin Franklin used glass in a new way to make bifocal spectacles). Over the next ten years, Amelung invested more money in glassmaking than had anyone in America before him. He produced large amounts of impressive table glass, much of it engraved (60). Yet he, too, failed, eleven years later. However, Amelung left us the best of early American glass, sometimes signed and dated.

An ironic contemporary view of the situation indicates that Amelung and others during the 18th century failed because they overestimated the market: "Most new works have been begun too large in this country....If we built a glasshouse, it was at the expense of Thousands and calculated to cover all that part of the country with glass which was not covered by houses."

Conjectural rendering of one of Amelung's glasshouses, drawn by Richard Stinely.

60. **Wheel engraved covered tumbler, inscribed, "Happy is he who is blessed with virtuous children. Carolina Lucia Amelung. 1788." It was made for glassmaker Johann Friederich Amelung's wife and illustrates the biblical story of the angel leading the child Tobias on a journey to cure his father's blindness.**

Despite many financial difficulties, problems stemming from the War of 1812, and competition with the English, American glass production increased during every decade of the 19th century. Most of the glass was still primarily for bottles or windows. Such tableware as was made in bottle and window glass factories was usually made by the individual worker for his family and friends, often on the workman's own time and at the end of his shift. This glass was frequently decorated with superimposed bits of glass and threads in the German tradition but according to the workman's whim (61, 62).

As settlers moved west of the Alleghenies, they constituted a new market. However, glass was difficult to ship overland and most western settlers did without it until the industry, too, moved westward. A traveler in the Western Reserve in the early 19th century commented: "The furniture for the table is...scanty and inconvenient...articles of crockery are few and indifferent. ...For want of a glass...from which to drink, if you are offered whiskey (which is the principal drink here) the bottle is presented to you or a bowl or a teacup containing the liquor...." Another traveler, Henry C. Knight, wrote in a letter, "They have no cider or common beverage for tabledrink; but, instead for the ladies water unqualified, and for the gentlemen, either whiskey, or apple or peach brandy...."

61. **Lily pad pitcher. The "lily pads" have been added and drawn up the sides of the pitcher. Possibly Lancaster or Lockport, New York, Glassworks, ca. 1840-1860.**

62. **Blown jug and sugar bowl with furnace-worked decoration, probably made on the worker's own time. Aquamarine glass is typical of New York State glass. These pieces are made with hollow bulbs in the stems and below the hen-shaped finial of the sugar bowl cover. The bulbs enclose silver half-dimes dated 1829 and 1835. Births, marriages, and other events were often commemorated by the insertion of coins with the appropriate dates. Ca. 1835-1850.**

In the 1790s, James O'Hara and Isaac Craig of Pittsburgh started a bottle manufactory (the first bourbon whiskey had just been distilled by a Baptist minister in the bluegrass country of Kentucky). A second factory was started south of Pittsburgh by financier Albert Gallatin, later co-founder of New York University and Secretary of the Treasury under Jefferson and Madison. In Pittsburgh, river transportation to the entire western frontier guaranteed a ready market. Nearby coal deposits provided ample fuel. By 1817, when President Monroe wanted American cut glass for the White House, Pittsburgh was the place from which to order it. From Pittsburgh, the industry spread down the Ohio River to western Virginia, Kentucky, and Ohio.

Skilled glassmakers were frequently on the move, seeking either better wages or the free land available in the western territories. This created a labor shortage and stimulated a need for greater productivity. Obviously, one way to speed production was to blow glass into a mold, as the Romans did, producing shape and surface pattern in one operation. It was an idea in use in

63. **Amber sugar bowl with mold-blown ribbing. Probably Zanesville, Ohio, ca. 1815-1830.**

64. **Tumblers with cut decoration and enclosed portraits of Jackson and Washington. Made at Benjamin Bakewell's Pittsburgh glasshouse, ca. 1824.**

England. The earliest molded American glass imitated cut glass. A housewife's book, published in New York in 1815, suggests:

> Those who wish for Trifle dishes, butter stands, &c. at a lower charge than cut glass may buy them in moulds, of which there is a great variety that looks extremely well if not placed near the more beautiful article.

Bottles and flasks for liquor were also mold blown; after the War of 1812, they were often embellished with patterns, scenes, Masonic symbols, the American eagle (65), and portraits of celebrities. George Washington's face appears on sixty-one different types of flasks. When Jenny Lind, the "Swedish nightingale," was brought to the United States by P.T. Barnum in 1850, several glasshouses were ready with bottles decorated by her portrait.

While the use of molds speeded production, it was the development of the mechanical pressing machine in the 1820s which made mass production of American tableware possible. In fact, the greatest contribution of America to glassmaking, and the most important development

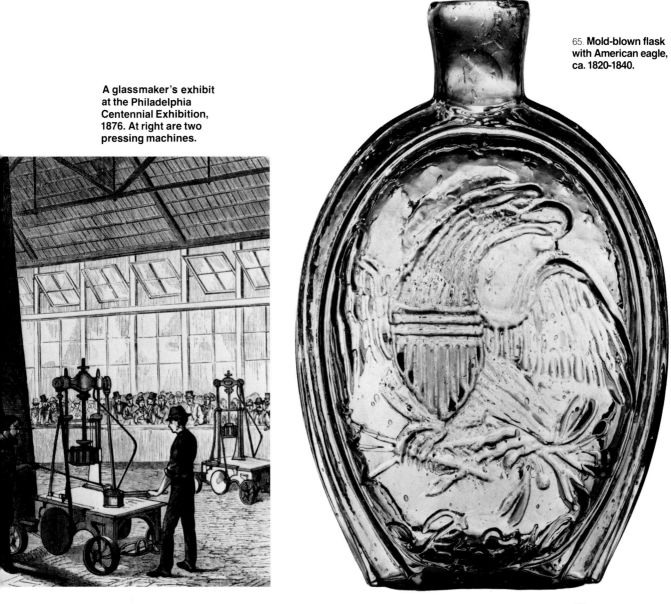

65. **Mold-blown flask with American eagle, ca. 1820-1840.**

A glassmaker's exhibit at the Philadelphia Centennial Exhibition, 1876. At right are two pressing machines.

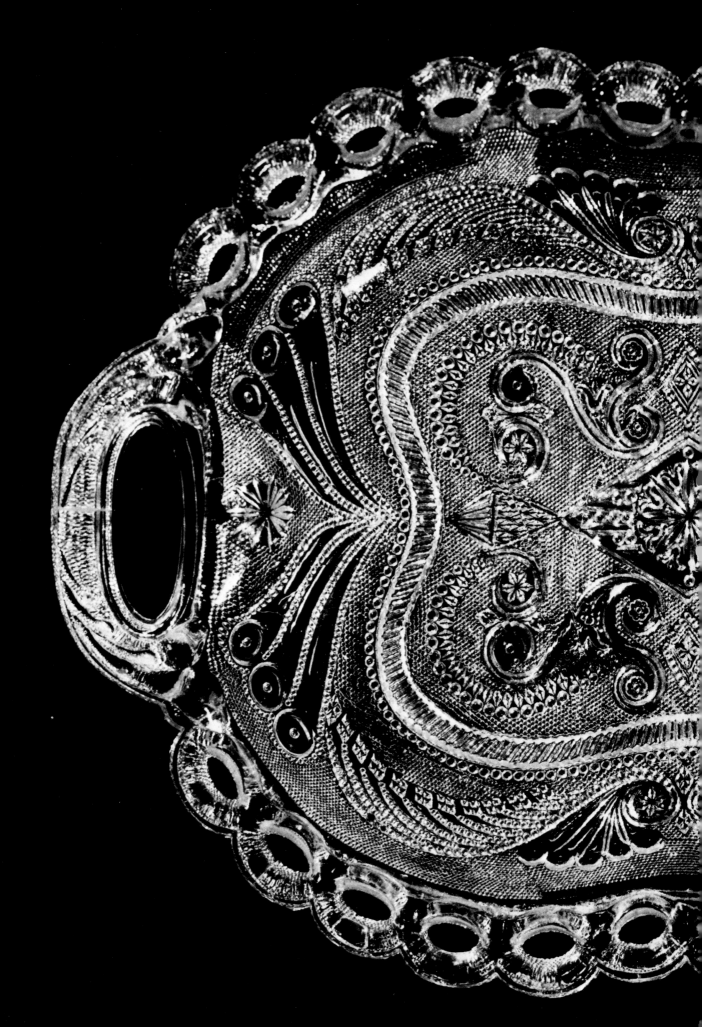

since the Romans learned glassblowing, was the sudden speed-up in the manufacturing process made possible by machine pressing. This development was a major turning point in the history of glassmaking.

At a pressing machine, two men with a minimum of experience could produce four times as much glass as a team of three or four trained glassblowers. A bowl came out of the press completely decorated in a few seconds. However, because contact of the hot glass with the cold mold produced a network of wrinkles on some pieces, moldmakers turned from imitating cut glass to lacy patterns (66) with a background of tiny dots; where every square inch was patterned, the wrinkles did not show. Technological innovations finally eliminated the wrinkles and patterns became simpler in the 1840s (67).

Pressing was used not only for candlesticks, vases, and table glass — facilitating the production of matched sets — but it made oil lamps (68) cheaper and more available to the general user. Whale oil was the most popular lamp fuel because it was abundant and gave what John Adams described as "the clearest and most beautiful flame of any substance that is known in nature." Whale oil lamps burned more brightly with glass chimneys, while the improved oil lamps invented by Argand employed both chimney and a glass shade to diffuse the light. At ten candlepower, these lamps were considered too harsh by the standards of the day. Count Rumford, a scientist writing in 1811, said, "No decayed beauty ought ever to expose her face to the direct rays of an Argand lamp," and doctors warned of eyestrain for users. (It would require five Argand lamps to equal the light provided by one modern 60 watt bulb.)

When kerosene lamps began to replace whale oil lamps, they not only used but *required* glass chimneys to burn properly and so a whole new glass industry to make them was born. Kerosene lamps, despite the introduction of gas lamps, were the major home lighting device until Edison developed the electric bulb in 1879. The first light bulb blanks were blown for Edison by Corning Glass Works the following year, but electricity did not supersede kerosene lamps in the countryside until rural electrification began toward the end of the Great Depression.

68. **Lamps, ca. 1840-1860.**

Opposite:
66. **Pressed cake plate with openwork rim. Probably the Boston & Sandwich Glass Company, Sandwich, Massachusetts, ca. 1830-1840.**

67. **"Comet" patterned glass. Made in both New England and the Pittsburgh area, ca. 1850. Gift in memory of Amy Chace.**

By the mid-19th century, American railroads and industry were growing and the country was rapidly expanding westward. Economic prosperity increased steadily; Mark Twain called the period after the Civil War "The Gilded Age." With prosperity came a taste for complex decoration and ornate styles, greatly influenced by fashions from abroad. Some glass imitated silver, tortoise shell, or Chinese porcelain. New shaded color effects were greatly admired, as were applied decoration and enameling. By the 1880s, American styles had exotic names such as "Amberina," "Burmese," and "Peachblow." Said the *Crockery and Glass Journal* (1886), "Just at present, art glass is all the go."

Patterns in glass cutting were also becoming more and more elaborate (69); a decanter and wineglasses made for the Philadelphia Centennial in 1876 were so profusely cut that not a half inch of surface was left undecorated. (At the Centennial exhibition, there was a huge glass fountain seventeen feet high, ornamented with cut crystal prisms, lighted by 120 gas jets, and surmounted by a glass figure of Liberty.)

Opposite:
70. **Flasks, jars, and bottles. Ca. 1840-1900.**

Glass containers (70) continued to be an important part of the American glass industry. In 1880, more than 25 per cent of the glass made in the United States was for liquor flasks, patent medicine bottles, and the home preserving jars perfected by John Landis Mason in 1858. These were all produced by blowing into molds and hand finishing the necks and lips. Patent medicines, promoted as cure-alls and often highly alcoholic, were a polite substitute for the liquor so frowned upon by the temperance movement. The bottles were usually distinctive and identified the contents: Turlington's Balsam of Life, Swain's Panacea, Perry Davis's Pain Killer, Lydia Pinkham's Vegetable Compound, and Hostetter's Celebrated Stomach Bitters. Just as in Roman times, glass was, and is, the preferred container material for many substances from food and beverages to scents, cosmetics, and panaceas.

In 1903, Michael Owens invented an automatic bottle blowing machine. For the first time, bottles could be made mechanically rather than by hand. This revolutionized the container industry. Later, other machines were perfected to make tableware, cookware, and fruit jars, putting more and more glass objects within the budget of most people. In 1926, the Wellsboro, Pennsylvania, plant of Corning Glass Works installed a ribbon machine, an automatic blowing device used primarily to manufacture light bulbs and, after 1939, Christmas tree ornaments. It can blow two thousand bulbs a minute, millions in a day.

69. **Cut glass dish with web of stars. Called the "Russian" pattern, it decorated a large service ordered for the White House in 1885. T.G. Hawkes & Co., Corning, N.Y., ca. 1900. Gift of T.G. Hawkes & Co.**

NINETEENTH
CENTURY
EUROPEAN GLASS

BIEDERMEIER GLASS

During the years between Napoleon's defeat at Waterloo in 1815 and the revolutions of 1848 in France, Austria, and elsewhere, the up-and-coming middle class in the Austro-Hungarian Empire—and especially in Vienna—enjoyed a prosperity and domesticity whose ideal has been described as "happiness in a quiet corner." It was the time of both Schubert and of the Viennese waltzes of Johann Strauss *père*, father of the composer of *The Blue Danube*. Peace, comfort, and leisure were among the most valued conditions of life, as was a well-ornamented home. The crafts which enhanced domestic life—furniture, silver, porcelain, glass—flourished.

Gottlieb or "Papa" Biedermeier, from whom the period takes its name, was not a real person but the fictitious author of comic verse which appeared regularly in a 19th century German magazine. He was intended as the personification of the Austrian bourgeois. *Bieder* in German means plain and inoffensive; *Meier* is a common German surname. When the term "Biedermeier" was first applied, it was used in a derogatory way. However, by the end of the 19th century, the Biedermeier style came to be greatly admired for craftsmanship and charm.

During the Biedermeier period, glassmaking was one of the most important industries in Austria, especially in Bohemia. The styles—despite the meaning of *bieder*—were hardly plain (72).

71. **Cut glass table. Fine glass furniture has enjoyed an intermittent vogue from the mid-18th century to the present day. This table was made of an octagon of blue glass above a spiral-cut amber glass column, the whole set upon a square amber glass base. Russia, the Imperial factory, probably a present from Czar Alexander I to his mother or sister, ca. 1804.**

72. **The Kulm Goblet —
elaborately cut, en-
graved, stained, and
enameled. Exemplify-
ing the essence of the
Biedermeier style, it is
one of three known
goblets made to com-
memorate the dedica-
tion of a monument to
the Battle of Kulm in
the Napoleonic War.
Bohemia, 1835-1850.**

By means of cutting, engraving, and enameling (73), glassmakers lavishly decorated their wares with views of Vienna, landscapes, allegories, flowers, animals, and other themes reflecting the romantic, comfortable atmosphere of the time. Glass engravers and enamelers reached a peak of skill in portraiture (74). Much Biedermeier glass was cased in various colors, some brilliant, others reminiscent of natural stones such as marble and porphyry.

VICTORIAN GLASS

An era which reveled in bustles, beaded reticules, fanciful furbelows, and Gothic cottages, in refinement, respectability, etiquette, and croquet, the Victorian Age was also the time when the British Empire was at its most prosperous and powerful, and the English middle class was coming into a position of dominance. Victorian glass reflected this prosperity and the accompanying love of elaboration, innovation, sentiment, and gentility. "The desire for decoration," said *The Decorator and Furnisher,* a periodical of the time, "amounts to a perfect craze, and no one can tell when it is going to end."

The lifting of the excise tax on glass in 1845 and the 1851 Great Exhibition of the Industry of All Nations, popularly known as the Crystal Palace Exhibition, in London further stimulated the glass industry — the Crystal Palace having its effect not only in Britain but all over the western world. Organized under the leadership of Prince Albert, consort to Queen Victoria, the world's first world's fair was held in a gigantic greenhouse-like building that covered almost twenty acres, using some four hundred tons of sheet glass, almost a million square feet. Three hundred thousand panes were hand blown at a single glasshouse and installed within six months, enclosing an area four times that of St. Peter's in

73. **Enameled tumbler with view of Meissen, by Samuel Mohn. He painted scenes of incredibly minute detail in translucent enamel: maps of celebrated battlefields, emblems of friendship, scenes of daily life, and views of cities. Germany, early 19th c.**

74. **Engraved portrait, signed D. Biman who was the foremost engraver of his era. Bohemia, 1834.**

Rome. No building on a comparable scale had ever been built, let alone covered with such a prodigious amount of glass: the Crystal Palace marked a revolutionary moment in the use of glass in architecture. Fifteen thousand exhibitors from nearly ninety countries gathered to show industrial art as well as handcrafts. Six million people visited it and it became the focal point of Victorian taste. Among the luxuriance of objects on display, glass was prominent; in fact, the centerpiece of the entire Palace was a complex glass fountain ("the gem of the transept") twenty-seven feet high. The Crystal Palace Exhibition was the first of a great number of influential international fairs, occurring nearly every year in the second half of the 19th century.

Large matching table services became more common and were all the rage. For example, the "Axel" pattern, manufactured in Sweden, offered eleven types of stemmed drinking glasses, plus a water beaker, beer tankard, various decanters, ewers, jugs, dishes, salad bowls with and without handles, cheese dishes, sugars and creamers, saltcellars, butter dishes, honey jars, tea caddies, flower vases, finger bowls, candlesticks, and bonbon dishes. To match were face-powder boxes, toilet-water flacons, soap dishes, water basins and jugs, and toothbrush holders—all of these matching items magnificently decorated with rows of glittering facets and fan-shaped edges. Many elegant houses could boast that the accessories of the boudoir matched those of the dinner table.

Styles as well as functional forms were proliferating at a dizzying rate. A catalog from one Stourbridge firm showed, on one page alone, fifty-six different styles of decanter—and offered 606 altogether.

The revival of earlier styles, such as Gothic and Renaissance, and the adaptation of "foreign" fashions, such as Egyptian and Moorish, typified 19th century taste. Glassmakers attempted to keep pace with the succession of "historical"

The Crystal Palace Exhibition.

75. Cameo plaque, carved by George Woodall from two layers of glass, one plum color and one white. As the white glass was worked thinner and thinner, it took on a bluish cast. There is no blue glass in this piece, which depicts Cupid and Venus by a pool. Stourbridge, England, ca. 1890.

76. Salamander paperweight. The salamander's body is of cased glass cut on the wheel to simulate scales. The legs and other details were added. Salamanders were long revered by glassmakers for it was believed that they could survive any fire. France, probably Pantin, ca. 1875. Gift of the Hon. Amory Houghton.

styles by copying earlier objects or inventing compatible forms to fill out services of glassware. One German lampworker adapted Venetian Renaissance designs so successfully that his glass has only recently been identified as being of 19th century date, not 16th century (77). Complicated centerpieces in colored glass were copiously supplied with contrasting threads and prunts. Especially in Stourbridge, so-called "fancy glass" came into vogue, featuring fluted edges, applied decoration, and a wide variety of color combinations.

77. **Covered dragon-stemmed goblet, so expertly made in the Venetian style that it was long regarded as a 16th century masterpiece. It was exposed as a 19th century fake in 1978. Hamburg workshop of C.H.F. Müller, ca. 1880.**

When glass pressing was introduced to England from America, the fantasy parade continued—opaque and colored glass ornaments assumed countless eccentric shapes: boats, shoes, swans, setting hens, wheelbarrows, and baskets. Many pieces imitated cut glass.

The larger the Victorian piece, the better; centerpieces and glass lighting fixtures reached monumental proportions. In fact, such was the Victorian demand for grandeur in glass that, in 1895, the Maharaja of Gwalior commissioned a chandelier which he desired to be greater than any in all of Buckingham Palace. When he was advised that the ceiling of his palace would not hold such a weight, he hoisted his largest elephant to it to prove the roof would hold.

Bohemian wheel engravers, tempted by the prosperity of Victorian England, emigrated to Stourbridge, and there made relief and intaglio pieces of great brilliance, often on the newly fashionable oriental themes. Perhaps the greatest of these Bohemians was William Fritsche, who eventually came to employ watery motifs particularly suitable to his polished engraving style. A Fritsche ewer (78) was explained with thoroughly Victorian sentiment:

> *The whole ewer may be said to represent the progress of a river from its birth in a rocky hillside until it loses itself at last in the blue infinity of the sea. The neck of the ewer represents the mountain birth-place of the stream.... The rush and hurry of the rapid river are wonderfully expressed by the strong, clear, curving volutes of the body.... The lowest part...is formed of a great fluted shell...symbolic of the bottom of the sea. In this swim and sport a circle of vigorous dolphins.*

The first European revival since Roman times of cameo-cut glass was yet another glory of Stourbridge, spurred by the work of John Northwood, who started in glassmaking at the age of twelve. Northwood later became famous for his adaptation in glass of the classic carvings from the Parthenon known as the Elgin marbles, and for his faithful glass copy of the famed Portland Vase, a Roman masterpiece of the 1st century B.C.

Cameo glass came in various colors, primarily blue or a plum color cased in white. A great variety of these works by Northwood, George Woodall (see page 76), and others—vases, dishes, plaques, scent flasks—were produced with classic themes. "It may be said with truth," proclaimed *The Pottery Gazette* of January 1, 1908, "that no more noble ornament can be conceived in a room than a fine well-designed and artistically executed cameo vase; beautiful in form, in colour and detail, and at the same time having a gem-like quality which is truly precious."

Around the middle of the 19th century, glass paperweights (see page 77) to hold down papers on a desk became popular. They developed at a time when writing paper was growing less expensive and letter writing was flourishing. Though they could be purchased at a modest price in stationery shops, paperweights soon came to embody all the most exquisite and bewitching technical effects of which glass is capable. Miniature, richly colored flowers, fruits, birds, reptiles, and commemorative portraits were encased in a heavy dome of clear glass which magnified the motif. Cut facets were often applied to the dome, creating both miniaturized, magnified, and multiplying images. Flowery and lacy effects were sometimes combined to produce delicate overall patterns. Small and fascinating, paperweights have come to be one of the most admired glass forms, sought by collectors all over the world.

78. **Engraved ewer by William Fritsche, who spent two and a half years working on it. Stourbridge, England, 1886.**

ONE HUNDRED YEARS OF MODERN GLASS

O f the many critical moments in the history of glass, five are of particular importance: first, the discovery of glass itself, more than 3500 years ago; second, the invention, about the time of Jesus, of glassblowing, so that glass became available in large quantities; third, the development of lead crystal with its greater weight and brilliance; fourth, the development in America of the mechanical press, so that glass objects could be produced at low cost. These were all technological improvements, each of which changed glass history.

The fifth turning point was of a different, nontechnical sort. It occurred about a hundred years ago, when designers and artists—who had not been glassmakers themselves—began to work in glass. Before then, glass design was usually dictated by the skill, taste, and ingenuity of the artisan who actually made the glass or the man who made the mold. By the 1870s, at least two artist-designers were working in glass, but it was not until the 20th century that the designer and artist became important in many of the world's best glasshouses.

This involvement of artists and designers in glass manufacture led to the emergence of "art glass," what the French call *vases de délectation*. Designers created designs to be made by others, that is, by craftsmen in a workshop or factory. Artists, among them painters, designed glass objects, too, and some also made the glass object itself at the furnace.

Among the first designer-artists to work in glass were Eugene Rousseau and Émile Gallé, both becoming known for their glass at the Paris Exhibition of

79. **Blown vase with applied and engraved decoration. Eugene Rousseau, France, 1884. Gift of George D. Macbeth.**

1878, one of the many international fairs which followed the 1851 Crystal Palace Exhibition.

Admiral Perry had opened Japan to trade with the West in 1854 and, by the time of the 1878 Exhibition, Japanese art was much admired in Europe. Rousseau (79), a designer, was greatly influenced by Japanese pottery and landscapes. His work was characterized by flamelike colored streaks and textures inlaid in the glass. Sometimes he added crackle effects and metal particles. The objects were decorated with enamel, cutting, and engraving. Rousseau derived his forms not only from oriental art but from the German *humpen* and from Italian Renaissance shapes. He made a significant contribution to glass design, but his career was short and his output small.

Émile Gallé is considered the mainspring of the Art Nouveau style in glass. During the 1880s, Art Nouveau was a graceful style widely used in most of the fine and decorative arts: architecture, paintings, posters, book illustrations, furniture, wallpaper, fabric, embroidery, jewelry, and glass. It developed at a time when intellectuals and artists, rebelling against the new industrialism, were glorifying both the craftsman and the high standards of the medieval guilds. Richly ornamental and often asymmetrical, it employed long, sinuous lines, weaving tendrils, and flowing rhythms. Art Nouveau was a style particularly suitable to the liquid, flowing qualities inherent in glass.

Like Rousseau, Gallé admired Japanese art, an admiration he expressed by soft, liquid colors, by themes from nature, and by a poetic mood.

To Gallé, nature was the source of all beauty. He stated this belief in a sign on his workshop door which read, "Our roots lie in the soil of the woods, in the moss by the rim of the pool." All of his life he was a practicing gardener and a learned writer on horticulture. He said that "each kind of plant possesses its ornamental style." He did not limit himself to elegant flowers like the orchids and lilies used as fashionable motifs at the time, but often used more homely plants like the thistle and the oak. His natural themes were enlivened by stylized insects, especially dragonflies (80).

80. **Bowl with colored and wheel engraved decoration. Amidst the dragonflies is the signature of Émile Gallé. France, ca. 1900. Bequest of Ellen D. Sharpe.**

Gallé was inspired by many of the symbolist poets and often decorated his vessels with ornamentally lettered quotations from Paul Verlaine, Charles Baudelaire, Stéphane Mallarmé, Victor Hugo, and Edgar Allan Poe, among others. The poetic quotations were always appropriate to the form and decoration of the vessels. Called *verreries parlantes*, these "talking glasses" were in the spirit of artists such as William Morris, Aubrey Beardsley, Dante Gabriel Rossetti, and — much earlier — William Blake, all of whom combined painting with poetry. Gallé dedicated one such vase to Sarah Bernhardt, the most celebrated actress of the time; it read, "'*de la lumière! de la lumière!* [Light! Light!]' Hamlet."

Gallé was able to make glass himself, but as his work grew in popularity, his establishment gradually developed into a large organization of craftsmen whom he carefully supervised. They carried out Gallé's designs in Gallé's personal, lyrical style. Gallé's glass was always signed, giving it an aura of art and adding value for collectors. Other glassmakers soon began to sign their wares, too. (Before Gallé, signed glass is rare, though we must not forget the Roman Ennion.) Many other factories, such as Daum in France, Val-Saint-Lambert in Belgium, and Kosta in Sweden, produced glass in the Gallé style.

When Louis Comfort Tiffany, an American painter, visited Paris in 1889, he saw Gallé's fantastically colored glass at the World Exhibition and was deeply impressed. Tiffany was the son of the founder of the famous New York jewelry store, Tiffany and Co. He, too, brought a fine arts background to the design of functional objects — and he, too, was affected by the art of Japan. His initial efforts in glass were in making stained glass windows, beginning in 1878. It was the leftovers from window glass production that Tiffany first used for blown glass. Like Gallé, he was a nature lover; unlike Gallé, he did not make glass at the furnace himself, but rather, he designed it. He was also a leading interior decorator with such eminent clients as the Union League Club in New York and Mark Twain, whose home in Hartford, Connecticut, he decorated. Tiffany was invited to redecorate the White House when Chester Alan Arthur was president.

Tiffany's first collection of blown glass was strongly expressive of Art Nouveau. It was exhibited at the Columbian Exhibition in Chicago in 1893 and was an immediate success. The metallic iridescence, its chief characteristic, was inspired, he said, by the iridescence resulting from decay on excavated Roman glass. Tiffany's glass had a silky, delicate patina over luminous colors. The metallic luster was a film of metal produced by exposing the glass to vapors or by direct application. The effect was sometimes heightened by etching the surface, creating a satinlike texture. It is said that $20 gold pieces were dissolved in acid and used as a source for the gold in the metal film.

Tiffany's blown forms were often sensual flowers on attenuated stems or other shapes which had not been used in glass containers before. These forms were decorated with lines and motifs which seemed to swell and narrow. One of his favorite patterns imitated a peacock feather. He also used many ancient techniques, such as millefiori, intaglio, and cameo carving. Tiffany named his glass *Favrile*; according to a brochure, the word was derived from the Old English word "fabrile," meaning "belonging to a craftsman or his craft." In Tiffany's catalog for the 1900 Paris Exhibition, he declares that "in none of the specimens of this glass is there any application of decoration by painting. Such designs as are found are in all cases produced by the combination of different colored glass during the operation of blowing the piece."

A French glassmaker, René Lalique, first became famous as a designer of Art Nouveau jewelry. Subsequently, he received a commission from the French perfumer Coty to produce some special perfume bottles for Coty's costly scents. He ultimately devoted himself completely to glass, which he called "an enchanted substance." The style most associated with his name is reflected in vases, bowls, and jars decorated in high, frosted relief produced by molding, sometimes accentuated by red or black enamel (82). Among his most frequent decorative motifs were the human figure, birds, fish, insects, and flowers, usually treated as elements in a formal pattern. Sometimes, the pattern was purely abstract. Sharp and icy, Lalique's style was the opposite of the flowing Tiffany glass and seemed at home in the Roaring Twenties, the age of the Charleston, the flapper, the first talking pictures, and sleek Art Deco interiors. Lalique's urbane style was closely related to the logical, sophisticated functionalist concept of beauty which stated, among other ideas, that materials should be used "honestly," according to their nature, that forms should be simple, and ornament geometric.

Maurice Marinot, who also believed in the functionalist definition of beauty, was originally a painter, one of the influential *fauves* or "wild beasts" along with such painters as Matisse, Roualt, Dufy, and Braque. In 1911, after visiting a glasshouse belonging to friends, Marinot became captivated by the qualities of glass as a material. He tried enameling on glass but soon turned to making the vessels himself; he was the first artist to undertake glassmaking single-handedly. His massive vessels were deeply cut, heavily etched, and decorated with clouds of bubbles and subtle color effects. Unlike Gallé and Tiffany, he did not draw his inspiration so much from nature as from the qualities inherent in the material and from the process of making it (83).

82. "Tourbillons," vase with black enamel decoration. *Tourbillons* are whirlwinds. René Lalique, France, ca. 1925.

83. Acid-etched vase. Maurice Marinot, France, 1934. Gift of Mlle. Florence Marinot.

81. **The Tiffany-Massier Lamp. The Tiffany glass shade was made by rolling a bubble of clear glass in chips of other colors of glass. The surface was sprayed with a metallic compound. The iridescent pottery base was made by Clement Massier, one of France's leading ceramicists. United States, 1895-1910.**

At the furnace, Marinot worked by himself or with one helper. Personally working at the furnace was essential to him; it was there that his ideas developed. "To be a glassman," he said, echoing the experience of glassmakers past, "is to blow the transparent stuff close to the blinding furnace, by the breath of your lips and the tools of your craft, to work in the roasting heat and the smoke, your eyes full of tears, your hands dirtied with coal dust and scorched. It is to produce an order of simple lines in the sensitive material by means of a rhythm which matches the life of the glass itself, so that in due course you may rediscover in its gleaming stillness that life of the human breath which will evoke living beauties."

Marinot gave up glassmaking in 1937 and returned to painting, but his work in glass had an impact on all the art glass which followed.

Another interpretation of functionalist esthetics, a style called Swedish Modern, came to prominence in household furnishings. It was, and is, characterized by strong, clean lines and a natural use of materials. In glass, it emphasized the substance's "frozen liquid" character. It was first developed at the Orrefors Glassworks, then a small, little known glasshouse in the forests of southern Sweden. In 1915 and 1917, Orrefors hired two painters, first Simon Gate, and then Edvard Hald, to design glass. This was the first time a glass company had hired artists. The style developed by Gate and Hald influenced glass design for many years. Hald was later to describe the role of the glass artist-designer thus: "An artist working in glass is primarily the director of a drama featuring the master and his colleagues plus a glowing lump of semi-molten glass...."

Orrefors, with the help of Gate and other designers, began making a new form of cased glass which they called *Graal* glass (in English, Grail glass). In *Graal* glass, colored relief decorations like those used by Gallé were covered with clear crystal, thus acquiring a smooth outer surface over what was now an inner relief. Under the direction of Gate and Hald (84), Orrefors also produced luxurious

84. **Covered urn. Designed by Edvard Hald, Orrefors, Sweden, 1927.**

85. **Vase engraved with Bacchus and Ariadne. Lotte Fink, J. & L. Lobmeyr, Austria, 1925.**

engraved works, some deeply carved and others shallowly engraved. Eventually, the Swedish style and the hiring of artists as designers were adopted by other Scandinavian factories such as Kosta and Boda.

At about the same time in Vienna, designers led by architect and designer Josef Hoffman were producing simple, functional wares based on geometric shapes, the forerunners of much glass made today. Among the firms for which Hoffman designed was J. & L. Lobmeyr, which made glass of a dignified classical style in Vienna for three generations. Stephen Rath, a Lobmeyr nephew, founded a branch of the company in 1918 which made glass engraved to the designs of artists. (See page 85.)

In 1903, a glassmaker named Frederick Carder founded Steuben Glass in Corning, New York. He had emigrated from Stourbridge, England, at the invitation of T. G. Hawkes & Company, an American firm noted for cut and engraved glass. Though Hawkes intended Carder to produce blanks for cutting and engraving, Carder immediately began to make a variety of lustrous glass in

86. **Intarsia vase. Inlaid with soft, overlapping colors in a technique similar to the** *Graal* **glass made at Orrefors in Sweden. Frederick Carder, Steuben Glass, United States, ca. 1925. Bequest of Gladys C. Welles.**

many colors and many styles, including Art Nouveau. In 1918, Corning Glass Works bought Steuben Glass from Carder, who continued to design for the company for many years (86). When Arthur A. Houghton, Jr., great-grandson of the founder of Corning Glass Works, became Steuben's president in 1933, he brought together architects and artists, including John Gates and Sidney Waugh, as designers. Corning scientists had just developed a new optical glass for lenses; it was of such unusual brilliance, purity, and workability that Houghton decided to use this material exclusively, thus ending Steuben's production of colored glass. Almost immediately, the Steuben design team established a new, distinctive style in a material free of flaws. Within a few years, Steuben glass was being exhibited at expositions, galleries, and major museums. The company has continued to use only its clear, highly refractive lead crystal (87) and to employ and commission designers and artists to produce new designs — functional and ornamental, abstract and representational. Steuben glass has been presented as state gifts by every United States president since Harry S. Truman.

87. **"Arcus." Peter Aldridge. Steuben Glass, United States, 1977.**

In Czechoslovakia, the great tradition of Bohemian engraving continued until World War II. After that war, the Bohemian technical tradition burgeoned into a sculptural style still evolving today (see pages 92-93) which has had an important effect on American glass. The sculptures are usually one-of-a-kind works made under the supervision of glass artists in factory time especially set aside by the Czech government. Special programs in Czech schools train students in glass design. Among the most important contemporary Czech glass artists are František Vízner, Bretislav Novák, Jr., Věra Lišková, Stanislav Libenský, Jaroslava Brychtová, the engraver Jiří Harcuba, and Pavel Hlava.

In Venice, the firm of Venini has employed many well-known designers, including Tapio Wirkkala of Finland. Marvin Lipofsky, an American glass artist and teacher, has also worked at Venini.

Until recently, most glass, including art glass, was made in factories. In the United States in 1962, Dominick Labino (88), a glass scientist and artist, developed a formula for a glass which would melt at a relatively low temperature. This made feasible the introduction of a small glass furnace, suitable for a workshop with limited space yet holding enough glass for the needs of a single glassblower. This small furnace permitted artists to experiment with glass in their own studios. At the same time, artist Harvey Littleton (89) and others began a series of workshops in glass at The Toledo Museum of Art. The glass

89. **"Four Square." Blown and cut forms. Harvey K. Littleton, United States, dated 1975.**

88. **Vase with colored enclosures. Two overlays of green glass rise up like the shoots of growing plants toward a zone of pinkish orange. Dominick Labino, United States, 1969.**

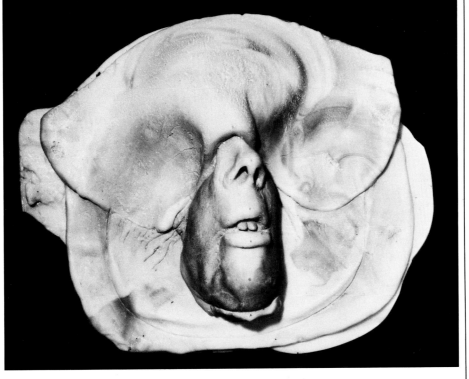

90."**Ravenna Grand Jury." Made in response to the shooting of students at Kent State University in 1970. Henry Halem, United States, dated 1972.**

91. **"Littleton the Teacher."** A portrait head of Harvey Little- ton, one of the found- ers of the studio movement in glass, by Erwin Eisch, a leading European glass artist. The head is one of a series, made by blow- ing a bubble of glass into a plaster mold and further shaping it after it was removed from the mold. The surface is enameled and the interior is silvered. Germany, 1976.

workshops, the new glass formula, and the small furnace combined to stimulate an international studio movement in which glass is used as an art medium.

Only about a hundred years have passed since Rousseau and Gallé began to make art glass; yet now this ancient substance is recognized as one of the materials of fine art, taking its place in the studio alongside paint and sculptor's clay and stone. Glassmaking is now taught in many art schools, craft schools, and universities. Sculpture executed in glass is displayed in museums and galleries. While some glass objects made in factories have always been nonutilitarian works created for ritual, ornamental, or esthetic purposes, today most works which are esthetically experimental or sculptural come directly from the artist's studio. In the past decade, a number of factories have recognized this and are increasingly employing studio artists.

In tracing the flow of glassmaking through history, we have seen glass begin in ancient times as a wondrous transformation, ordinary sand turned into a precious substance. This substance was regarded with awe as a magical manifestation and glassmaking's secrets were jealously guarded. In modern times, glass has retained its traditional fascination—the magic of its ability to capture and reflect light and of its solid transparency, being there and not

92. **Sculpture. Layered, laminated, and blown. Tom Patti, United States, 1978.**

there at the same time—as the French writer Marcel Proust said of glass in *Remembrance of Things Past:*

> *...real without being actual, ideal without being abstract...*

Now, toward the end of the 20th century, more than three millenia after its discovery, glass seems only to have begun revealing its magic and its marvels. New forms of this shiny, hard, fragile material are enabling man to journey not only to the frontiers of science in the laboratory and in industry, but perhaps to the beginning of our universe. In the 1980s, the American space shuttle—a vehicle partly composed of glass—is expected to put into orbit a telescope whose glass mirror will be capable of capturing light from suns 14 billion light years away. Since some scientists believe our universe was formed about 14 billion years ago, this glass mirror may provide a view of the creation of the first stars.

In the same era when glass is taking us to the very edge of outer space, it has become a medium for exploring inner space as well—an art medium with which the glass artist may transform his inner vision into a statement of universal meaning expressed in terms of transparency and light. In the coming decades, glass may well reveal not only more about our world but, as art must, more about ourselves.

93. **"Crystal Glass Sculpture," by Pavel Hlava, whose work has exerted a major influence on contemporary glass artists. Czechoslovakia, dated 1978.**

Opposite:

94. **Sculpture, made by Stanislav Libenský and Jaroslava Brychtová for the entrance lobby of the new Corning Museum of Glass. Czechoslovakia, 1980.**

BIBLIOGRAPHY

While any errors are the sole responsibility of the author, she made liberal use of the following admirable and authoritative books, all of which she heartily recommends to the reader who wishes further to pursue the history of glass.

Charleston, Robert J. *Masterpieces of Glass. A World History from the Corning Museum of Glass.* New York: Harry N. Abrams, Inc., 1980.

Corning Museum of Glass, The. *Glass from the Corning Museum of Glass: Guide to the Collections.* Corning, New York: Museum, 1958; revised, 1965, 1974.

——————. *New Glass, A Worldwide Survey.* Corning, New York: Museum, 1979.

——————. *A Survey of Glassmaking from Ancient Egypt to the Present.* Chicago: University of Chicago Press, 1977. Text-Fiche.

Newman, Harold. *An Illustrated Dictionary of Glass.* With an introductory survey of glassmaking by Robert J. Charleston. London: Thames and Hudson, 1977.

Polak, Ada. Glass: *Its Tradition and Its Makers.* New York: G.P. Putnam's Sons, 1975.

——————. *Modern Glass.* New York: Thomas Yoseloff, 1962.

Spillman, Jane Shadel. *Glassmaking: America's First Industry.* Corning, New York: Corning Museum of Glass, 1976.

Weiss, Gustav. *The Book of Glass.* trans. Janet Seligman. New York: Praeger, 1971.

An extensive reading list is available upon request from The Corning Museum of Glass.

OBJECT INDEX

21. Wheel abraded bottle.
H. 18.4cm (62.1.31)

22. Snake-thread beaker.
H. 11.5cm (72.1.5)

23. Frankish cone beaker.
H. 23.5cm (66.1.247)

24. Claw beaker.
H. 17.7cm (70.1.46)

25. Sasanian bowl.
D. (top) 8.1cm (72.1.21)

26. Octagonal pilgrim bottle.
H. 11cm (50.1.34)

Glass in the Far East

27. Pi disk.
D. 16.4cm (51.6.548)

28. Red-cased vase.
H. 48.9cm (57.6.10)

29. Five snuff bottles.
H. (of tallest) 9.5cm (51.6.309,
51.6.326, 57.6.2, 51.1.295,
51.6.338)

30. Enameled Ch'ien Lung
vases. H. 16.5cm (53.6.1)

Islamic Glass

31. Relief-cut bowl.
D. 14cm (53.1.109)

32. Relief-cut bottle.
H. 16.6cm (71.1.7)

33. Hedwig beaker.
H. 8.6cm (67.1.11)

34. Mold-blown bottle.
H. 25.8 (55.1.6)

35. Enameled and gilt vase.
H. 30.2cm (55.1.36)

36. Mosque lamp.
H. 30.5cm (52.1.86)

European Glass

37. *Cristallo* goblet.
H. 23.6cm (53.3.38)

38. Dragon-stemmed goblet.
H. 26.2cm (51.3.118)

39. Covered goblet.
H. (with cover) 34.9cm (64.3.9)

40. Wineglass.
H. 17.8cm (61.3.135)

41. Covered *cristallo* beaker.
H. (with cover) 29cm (50.3.1)

42. Spanish *cantir.*
H. 40.8cm (54.3.143)

43. *Lattimo* bowl. H. 5.9cm
D. (rim) 14.1cm (76.3.17)

44. Dragon-stemmed goblet.
H. 32.7cm (59.3.20) Covered
goblet. H. 46cm (65.3.56)

45. *Waldglas* beaker.
H. 20.2cm (53.3.2)

46. *Roemer.*
H. 27.8cm (64.3.92)

47. Enameled *humpen.*
H. 26.4cm (60.3.4)

48. Diamond engraved dish.
D. 48.8cm, H. 6.2cm, (77.3.34)

49. Maximilian *Pokal* .
H. 34.9cm (79.3.158)

50. Wheel engraved plaque.
H. 15.3cm, W. 11.5cm (76.3.29)

51. Covered chalk glass
beaker. H. (with cover) 21.9cm
(58.3.185)

52. Ravenscroft *roemer.*
H. 18.8cm (50.2.2)

53. Williamite political glass.
H. 19.2cm (79.2.31)

54. Diamond stippled goblet.
H. 25cm (50.2.10)

55. Enameled and gilt
goblets. H. 22.3cm (50.2.8)

56. Candelabrum. H. 90.5cm
(50.2.23)

57. Cut glass fruit bowl.
H. (ends) 21.2 cm (50.2.41)

American Glass

58. Covered sugar bowl.
H. 15.5cm (50.4.2)

59. Sugar bowl.
H. 15.5cm (50.4.18)

60. Wheel engraved covered
tumbler. H. 30.1cm (55.4.37)

61. Lily pad pitcher.
H. 18.5cm (50.4.450)

62. Blown jug and sugar
bowl. D. (rim) 10.9cm, H.
16.4cm (55.4.157, 55.4.131)

63. Sugar bowl.
D. 13cm (55.4.70)

64. Tumblers.
H. 8.5cm (55.4.57, 55.4.273)

65. Mold-blown flask.
H. 16.6cm (50.4.324)

66. Pressed cake plate.
L. 30cm (68.4.406)

67. Comet patterned glass.
H. (of tallest) 34.8cm
(68.4.12-23)

68. Lamps.
H. (of tallest) 31.1cm (68.4.375)

69. Cut glass dish.
D. 34cm (51.4.536)

70. Flasks, jars, and bottles.
H. (of tallest) 29.6cm

19th Century European
Glass

71. Cut glass table.
H. 79cm (74.3.129)

72. The Kulm Goblet.
H. 25.6cm (75.3.91)

73. Enameled tumbler.
H. 10.2cm (51.3.198)

74. Engraved portrait.
D. 9.5cm (65.3.68)

75. Cameo plaque.
D. 46cm (65.2.19)

76. Salamander paperweight.
D. 11.5 (55.3.79)

77. Covered dragon-stemmed
goblet. H. 35.6cm (51.3.115)

78. Engraved ewer.
H. 38.5cm (54.2.16)

One Hundred Years of
Modern Glass

79. Blown vase.
H. 9.2cm (62.3.112)

80. Bowl.
D. (rim) 12.4cm (54.3.30)

81. The Tiffany-Massier
Lamp. H. 81.5cm (79.4.117)

82. "Tourbillons" vase.
H. 20.1cm (74.3.27)

83. Acid-etched vase.
H. 17cm (65.3.48)

84. Covered urn.
H. 26.2cm (68.3.16)

85. Engraved vase.
H. 17.6cm (72.3.10)

86. Intarsia vase.
H. 17.5cm (69.4.221)

87. "Arcus."
H. 8.9cm, W. and D. 12.7cm

88. "Four Square."
H. 13.4cm W. (at widest point)
24.6cm (75.4.54)

89. Vase with colored
enclosures.
H.11.2 (70.4.24)

90. "Ravenna Grand Jury."
10.8 x 37.9 x 28.1cm (73.4.51)

91. Sculpture.
H. 14.1cm (79.4.3)

92. "Littleton the Teacher."
H. 50.3cm (76.3.32 C)

93. Sculpture.
H. (without base) 45cm, L.
53cm (78.3.50)

94. Glass Sculpture.
H. 305cm